SCIENTIFIC PROGRESS

SCIENTIFIC PROGRESS

*A Study Concerning the Nature of the Relation
Between Successive Scientific Theories*

FOURTH EDITION

by

CRAIG DILWORTH
Department of Philosophy, Uppsala University

 Springer

Author
Craig Dilworth
Kristinebergs Strand 3
SE-112 52 Stockholm
Sweden

ISBN: 978-1-4020-9108-7   e-ISBN: 978-1-4020-6354-1

Library of Congress Control Number: 2008934469

Printed on acid-free paper

9 8 7 6 5 4 3 2 1

springer.com

"Free of unnecessary ballast, and written with didactical aptitude, this book gives a complete overview of how the different views of scientific progress have developed since the time of the Vienna Circle. It is a suitable introduction to a complex period in contemporary theory of knowledge. In later chapters the author present his own standpoint, so that the work can also be used as a source of new impulses in this direction. ...

"The author convincingly works out how from his point of view it is possible to explain the conflict between two theories as an incompatibility of perspectives, and at the same time avoid sliding into relativism by giving criteria for scientific progress. ...

"I hope that my all too brief remarks will encourage the reader – and especially the interested non-specialist – to read this book."

Dialectica

"This book provides an extremely clear description and critique of the best known contemporary versions of philosophy of science, and a very suggestive ... solution of the general problem of scientific progress."

Annals of Science

"The views discussed are carefully referenced and traced back to original sources. In this respect the work is especially useful to anyone interested in general problems in the philosophy of science."

Choice

"Clear, interesting, and historiographically sensitive."

ISIS

"The topic is an exceptionally difficult, but extremely important one. Most of Dilworth's discussion is clear, well-written and technically flawless. Complemented by the right materials it should serve as a useful text for graduate and advanced undergraduate study."

British Journal for the Philosophy of Science

"Dilworth's work is clear and suggestive. The basic theses are presented with elegant philosophical sobriety, and the work as a whole can be called scientific not only for its subject matter, but also for its method."

Investigacion y Ciencia

"[The book] gives valuable instruction designed to keep one abreast of developments in philosophical reasoning."

Methodology and Science

This study differs from the stance commonly taken by epistemologists. The author has, for natural reasons, begun with the present state of the subject; he moves by degrees however to a position which is not only theoretically original, but which brings to a discussion that has become asphyxiated the oxygen necessary for it to regain its original epistemological content."

Epistemologia

"The book will quickly recommend itself, and reward the reader."

Aslib

"One of the most interesting contemporary approaches to questions related to the dynamics of science."

Revista de Filosofie

"An insightful and original work."

Risto Hilpinen, *University of Miami*

"This work must be considered one of the most significant contributions to appear in the present debate concerning the problem of scientific change and scientific progress."

Evandro Agazzi, *University of Genoa*

CONTENTS

APPENDICES

ACKNOWLEDGEMENTS

This study has been presented in a number of versions since its central ideas first appeared in a short essay 'Incommensurability and Scientific Progress' in 1975. Since that time it has received the helpful criticism of very many people, and to all of them I here express my thanks. Lennart Nordenfelt has discussed with me in detail versions appearing in 1976 and 1977. Those whose contributions are more recent include Rainer Carls, Paul Feyerabend, Mats Furberg, Lars Hertzberg and Dag Prawitz, each of whom has made valuable suggestions concerning my 1978 paper 'On the Nature of the Relation Between Successive Scientific Theories,' from which the last six chapters of the present work have been developed. More specialised comments have been offered by Staffan Nilsson (Chapter 10), and by Peter Gärdenfors and Włodzimierz Rabinowicz (Chapter 11). And for reading and commenting on the whole of this work just before it went to press, I express my gratitude to Ingvar Johansson and Giovanni Sommaruga.

Very special thanks are due to Prof. Stig Kanger, for his support and guidance during my years as a graduate student at Uppsala, and to Prof. Evandro Agazzi, who has done everything possible to help see this work through to completion.

FRIBOURG
March 1981   C. D.

PREFACE TO THE SECOND EDITION

This edition has been supplemented by two appendices. The first, which has also appeared as an independent essay in *Erkenntnis* **21** (1984), is a deepening of the discussion of theoretical terms which runs through Chapters 4, 10, and 11 of the book. The second is a paper written in response to a number of comments on the book, and focuses on the relation between the Gestalt Model and the Perspectivist conception of science. Its content slightly overlaps that of the main text, but this repetition may itself be of some use in that it involves the presentation of the central ideas of the book in a somewhat different form. This paper was read at the History and Philosophy of Science Conference in Veszprém, Hungary in 1984; and it will also appear separately in a forthcoming volume of the *Boston Studies in the Philosophy of Science* devoted to the proceedings of that conference.

Apart from these appendices, this edition also includes some minor corrections and a number of typographical improvements.

STOCKHOLM
January 1986   C. D.

PREFACE TO THE THIRD EDITION

This edition incorporates four new appendices (III–VI). The first was written shortly after the completion of the first edition of the book, and has appeared in Swedish in *Vår Lösen* **72** (1981). Intended for an audience of nuclear physicists, it includes the application of the Perspectivist conception to the phenomenon of nuclear modelling in physics. Appendix IV is a further development of the discussions in Chapter 10 and Appendices I and III on the nature of laws and theories, and has previously been published in *Zeitschrift für allgemeine Wissenschaftstheorie* **XX** (1989). Appendix V was written in response to a challenge to apply the Perspectivist conception to the Mach-Newton controversy concerning the nature of space, and has appeared in the anthology *Changing Positions*, Uppsala, 1986. And the last appendix, which has just been completed, involves the application of the Perspectivist conception to a central issue in ecological economics concerning the attainment of development in an environmentally sustainable way. An expanded version of Appendix VI will be appearing in *Population and Environment* **15** (1993–1994). Funding for research on this topic was provided by the Swedish Environmental Protection Agency.

As in the previous edition, minor changes and typographical improvements have also been made.

STOCKHOLM
November 1993

C. D.

PREFACE TO THE FOURTH EDITION

It is of course gratifying to see this book come out in another edition more than twenty-five years after its first publication, thereby bringing it to the attention of a potential new readership. On the other hand it is disheartening that since the book's first publication it seems that very few of my academic colleagues have been able to benefit from it, or, for that matter, even to understand it. That this is so is ironic, considering the many remarks my reviewers have made as to the clarity of the book's presentation. What I have come to realise, however, is that this inability stems at least in part from the novelty of the view I am presenting: that in order to understand it professional philosophers of science simply have had to take too great a step outside the logico-linguistic framework in which they have been educated. Confounding this shortcoming, perhaps abetted by the influence of this same framework, is the apparent inability of the majority of my commentators to appreciate the nature of philosophical theorising, such that they have been unable to distinguish it from the provision of a general description or analysis, and realise that what is presented in this book is in fact a theory.

For this edition of the work, apart from removing some repetitive text from Appendices III, V and VI, I have replaced the original Appendix II, and added two other new appendices. The original Appendix II was intended to function as both an alternative presentation of the book's central ideas and a reply to criticism of the first edition, while its replacement is intended only to perform the latter function, and this with direct reference to my critics. A similar reply to criticism of the second and third editions of the book may be found in Appendix III of the second edition of my other major work in the philosophy of science, *The Metaphysics of Science*.

The first of the other new appendices, Appendix VII, takes a look at the nature of the empirical aspect of science in different terms than those of its treatment in Appendices I and IV, and from a more historical point of view. And the second, Appendix VIII, presents a theory that, like the Perspectivist conception of science, is intuitively based on the Gestalt Model, but which concerns rather the topic of identity and reference in the philosophy of language.

This edition of the work has been completely reset; and, apart from the changes mentioned above, there have been many typographical and other minor improvements.

STOCKHOLM
April 2007   C. D.

INTRODUCTION

For the philosopher interested in the idea of objective knowledge of the real world, the nature of science is of special importance, for science, and more particularly physics, is today considered to be paradigmatic in its affording of such knowledge. And no understanding of science is complete until it includes an appreciation of the nature of the relation between successive scientific theories – that is, until it includes a conception of scientific progress.

Now it might be suggested by some that there are a variety of ways in which science progresses, or that there are a number of different notions of scientific progress, not all of which concern the relation between successive scientific theories. For example, it may be thought that science progresses through the application of scientific method to areas where it has not previously been applied, or, through the development of individual theories. However, it is here suggested that the application of the methods of science to new areas does not concern forward progress so much as lateral expansion, and that the provision of a conception of how individual theories develop would lack the generality expected of an account concerning the progress of science itself.

In considering the nature of scientific progress through theory change, a particular feature of the relation between theories presents itself as requiring explanation. This feature is the competition or rivalry that exists between successive theories in their attempts to explain certain aspects of reality. We note then that an adequate account of scientific progress should include a conception of the conflict that arises in the case of successive scientific theories.

In its treatment of the notion of scientific progress, this study begins with a critical analysis of the logical empiricist and Popperian conceptions of the nature of the relation between successive theories, and the basis from which these conceptions are derived. The analysis is structured via the reconstruction of the empiricist and Popperian conceptions in terms of the Deductive Model, which is formally identical to the covering-law model of explanation. This reconstruction is intended to show in detail what the empiricist and Popperian views

consist in, and in so doing demonstrate how they are in fact conceptually dependent on the Deductive Model, which thereby determines
both their capabilities and their limitations.

The major criticisms based on the reconstruction are of three
kinds. The first of these concerns the inability of the Deductive
Model, as employed by the empiricists and Popper respectively, to
formulate conceptions of important aspects of science. The most
important of these criticisms are that the empiricist view affords no
notion of theory conflict, and that the Popperian view fails to provide
a notion of scientific progress. The second type of criticism concerns
the problem of *applying* the Deductive Model to actual science.
Thus, for example, it is shown that while the empiricists have given
us a notion of scientific progress (as involving deductive subsumption), actual scientific advance does not take this form. The third
kind of criticism, directed mainly at Popper, suggests that a number
of claims considered to be integral to his view are actually quite
ad hoc, in that they are not at all suggested by the Deductive Model,
on which his conception of science depends.

The study then moves on to consider the important claims of
Thomas Kuhn and Paul Feyerabend, that in certain cases succeeding
theories might well be 'incommensurable' with their predecessors.
These claims, in their negative aspect, are viewed as essentially
being criticisms of the second sort mentioned above. That is, they are
seen to concern the applicability of the empiricist and Popperian
conceptions of the relation between theories, and to suggest the
relinquishment of the model underlying these views. And, in their
positive aspect, they are taken to suggest that in actual science theories are often related in the same sort of way as are the different
aspects of a gestalt-switch diagram.

Following this lead, a model which is fundamentally different from
that of the empiricists and Popperians is introduced. This model – the
Gestalt Model – is intended to provide a positive intuitive understanding of incommensurability, and to afford notions both of conflict and of progress.

Taking the Gestalt Model as an intuitive basis, the Perspectivist
conception of science and scientific progress – which constitutes the
heart of the book – is then presented. On the Perspectivist conception,
scientific theories are seen not to be entities of the sort which are

either true or false, but to be structures which are more or less applicable depending on the results of certain measurements.

Following this, the Perspectivist conception is developed further in the context of its application to the kinetic theory of gases. In this development the role of *models* in theoretical science, while not treated by the empiricist and Popperian views, becomes of central importance.

A critique is then made of the set-theoretic or structuralist conception of science, in which a notion of model also plays a role. An examination of the reconstruction of Newtonian particle mechanics in terms of intuitive set theory, and the attempted extension of these methods to the case of theory change, finds them not to have provided adequate conceptions either of theory conflict or of scientific progress.

Finally, the Perspectivist conception is applied to the views of Newton, Kepler, and Galileo concerning the motions of material bodies. Here the opportunity is also taken to compare the Perspectivist view with its alternatives, thereby further demonstrating its relative superiority.

CHAPTER 1

THE DEDUCTIVE MODEL

As will be shown in this study, the Deductive Model constitutes the formal basis upon which both the logical empiricist and Popperian conceptions of science and scientific progress are built. It is here introduced in its most familiar form: as a model of explanation and prediction.[1]

1. THE DEDUCTIVE MODEL AS A MODEL OF EXPLANATION

Popper formulates the model in his *Logic of Scientific Discovery*[2] as a model of causal explanation consisting in universal or general statements and singular statements, the conjunction of which entails some particular prediction. The model takes its more classic form as the covering-law model of deductive-nomological explanation in an article by Carl Hempel and Paul Oppenheim entitled 'Studies in the Logic of Explanation' (1948). It is there summarised in the following schema:

$$(1)$$

$$\left.\begin{array}{ll} L_1, L_2, ..., L_r & \text{General Laws} \\ \\ C_1, C_2, ..., C_k & \text{Statements of antecedent conditions} \end{array}\right\} \text{Explanans}$$

Logical deduction

$$\left.\begin{array}{ll} E & \text{Description of the empirical phenomenon to be explained} \end{array}\right\} \text{Explanandum}$$

---

[1] While the Deductive Model – which is being made explicit for the first time in this book – has the same logical form as the covering-law model, and may be seen as first having been expressed as that model, its function is different and its scope is greater. Where the covering-law model is to function as a model of explanation and prediction, the Deductive Model, which includes the covering-law model, is to afford a template for the reconstruction of the whole of the empiricist and Popperian philosophies of science.

[2] Popper (1934), pp. 59f.

Taken to its simplest extreme, the model may be schematised as follows:[3]

(2) $A \vdash B$,

where "A" is to denote the conjunction of general laws and statements of antecedent or initial conditions, and "B" is to denote the explanandum. The deductive relation is to go from A to B so that the truth of A is sufficient for the truth of B, and the truth of B is necessary for that of A.

The model may also be presented in a slightly more complicated form thus:

(3) $L \wedge C \vdash E$.

Here "L" denotes the conjunction of general laws, "C" the conjunction of statements of initial conditions, and "E" the explanandum.

In the model general laws are taken to be unrestricted universal statements; and statements of initial conditions and the explanandum may be conceived as restricted or specific statements. (Unrestricted statements are to be applicable at any place at any time, while restricted statements are to relate only to specific times and places. For example, on this line of thinking 'All swans are white' and 'There exists a black swan' are unrestricted; and 'This swan is black' is restricted.)

The formally deductive nature of the model can be captured by its being formulated directly in terms of the first-order predicate calculus. Here only one law and one statement of initial conditions will be taken to be present:

(4) L $\forall x\, (Fx \rightarrow Gx)$ (unrestricted)
     C $\underline{Fa\qquad\qquad}$ (restricted)
thus, E Ga (restricted).

And an example of the application of the above might go as follows:

(5) $\forall x\, (Fx \rightarrow Gx)$ All copper expands when heated
     $\underline{Fa\qquad\qquad}$ This is copper being heated
thus, Ga This (copper) expands.

---

[3] Cf. ibid., p. 76.

Here "*F*" stands for: 'is copper being heated,' and "*G*" stands for: 'expands.'

It is of some interest to note that the model, as given in (4) and (5) above, not only bears a close affinity to the Aristotelian syllogism,[4] but is in fact a Stoic syllogism of the same form as: All men are mortal; Socrates is a man; thus, Socrates is mortal.[5]

As presented by Popper and by Hempel and Oppenheim the model can function in two different ways: it can serve either as a model of explanation or as a model of prediction. In applying it as a model of explanation it is supposed that those receiving the explanation are aware of the truth of E, and are being informed that L and C are the case. In its application to prediction, L is assumed true, and, following the establishment of C, the truth-value of E is to be empirically determined.[6]

2. A CRITICISM OF THE MODEL AS A MODEL OF EXPLANATION

As presented above, the Deductive Model is to afford the linguistic form that a line of reasoning ought ideally to have in order to count as an explanation or prediction. In this guise it has been the recipient of a number of criticisms, most of which concern either the existence of seemingly adequate explanations (or predictions) not having the form suggested by the model (e.g. teleological explanations), or the fact that certain lines of reasoning have the deductive form, and yet are not explanations (e.g. conventional generalisations).

All the same, it is believed by many that the rigorous explanations made in deterministic branches of science do in fact have the deductive form, and that in this way it constitutes a sort of ideal, deserving of emulation. But it is suggested here that explanations in science do not have this form, and that those cases to which the model actually has been applied do not constitute instances of explanation.

---

[4] Cf. *An. Pr.* 26ᵃ 24–27: "Let all *B* be *A* and some *C* be *B*. Then if 'predicated of all' means [that no instance of the subject can be found of which *B* cannot be asserted], it is necessary that some *C* is *A*. ... So there will be a perfect syllogism."

[5] Concerning the Stoic origin of syllogisms having this form, see Bocheński (1961), p. 232&n.

[6] Cf. e.g. Hempel (1962), pp. 118–119.

If we consider the (paradigmatic) example given above, and suppose someone to witness the expansion of some particular material under certain conditions and to seek a scientific explanation of this phenomenon, it is not at all clear that his being told that the material is copper being heated, and that all copper expands when heated, would provide him with what he is seeking. In other words, his knowing that *all* copper behaves in this way under these conditions does not tell him why this particular piece does so; it only tells him that, in being constituted of copper, if this piece were replaced with another, also constituted of copper, the substituted piece would behave in the same way. And this would still leave him without an explanation as to why this material, which he now knows to be copper being heated, should expand under these circumstances.[7]

This problem will be seen later in this study to stem from the fact that explanations are here viewed as being based on scientific laws, rather than on theories. But of greater interest at this point is the fact that, as will now be shown, not only can both the empiricist and Popperian conceptions of science and scientific progress be derived from the Deductive Model, but both their capabilities and limitations are bound to it.[8]

---

[7] For a similar criticism, see Scriven (1962), p. 203; concerning the applicability of the Deductive Model to the case of laws being explained by higher-level theories, see Chapter 4 below.

[8] The present reconstruction of the logical empiricist and Popperian conceptions in terms of the Deductive Model may be seen as a presentation of what has recently come to be called 'the statement view.' Cf. e.g. Stegmüller (1973), p. 2, and Feyerabend (1977), p. 351.

CHAPTER 2

THE BASIS OF THE LOGICAL EMPIRICIST CONCEPTION OF SCIENCE

1. VERIFIABILITY

Logical empiricism is an outgrowth of logical positivism, in which the verifiability principle was put forward as a criterion for distinguishing meaningful statements from meaningless pseudo-statements. For logical positivism, if any proposition or statement were not in principle conclusively verifiable by experience, it was to be considered meaningless, or, at best, tautological. Along this line then it was intended that meaningful statements include the pronouncements of science, while excluding those of metaphysics, ethics, and theology.

With the realisation that on this criterion scientific laws would themselves be meaningless, a first step towards the logical empiricist position was taken by extending the status of meaningfulness to any proposition from which an empirically verifiable proposition could be logically derived. In such derivations, the meaningfulness of the consequent was to imply that of the antecedent.

But this view too suffered problems, a major one of which centres on the fact that no universal statement or law by itself entails an observation statement. What more is required is a statement of the conditions under which the observation is being made.

In this way then alterations in the logical positivist criterion of meaningfulness give rise to the basis of the logical empiricist conception of science, in which scientific laws, in conjunction with statements of initial conditions, are to entail particular observation statements. Here we see that lying behind these developments is the conception of laws, statements of conditions, and observation statements as having the form suggested by the Deductive Model: $(L \wedge C) \vdash E$.

Unfortunately, for logical reasons, the above attempt to include scientific laws among meaningful assertions while excluding metaphysical sorts of statements has proved unsuccessful.[1] But more important here is the introduction of the Deductive Model at the basis of the empiricist conception of science.

2. INDUCTION AND CONFIRMATION

As well as affording a structure for explanation and prediction, and for the above criterion of meaningfulness, the Deductive Model can be seen to form the basis of the empiricist conception of induction. Where in applying the model to explanation the starting point taken is the truth of the explanandum, and in the case of prediction it is the truth of the laws and statements of conditions, in its application to induction one takes the statements of conditions and explanandum, i.e. $(C \land E)$, or $(Fa \land Ga)$, to be true. Thus while we should not say that scientific laws, as conceived on the model, are logically derivable from statements of conditions and explananda, we may say that they can be related to the latter by means of induction.

A point not always recognised in discussions concerning (empirical) induction is that the term has two distinct applications. In one, induction may be thought of as a possible means by which we come to realise that there exist certain regularities in nature on the basis of an acquaintance with their instances. In the other, induction may be considered the method employed to afford rational support for the claim that some particular regularity does in fact exist. As conceived on the Deductive Model, both of these applications are fundamental to logical empiricism, the latter (called 'confirmation') defining its position in the context of justification, and the former (here termed simply 'induction') in the context of discovery.

The main problem with the empiricist conception of *induction* as being the means by which new laws are discovered, as has been noted by others, is that it provides no hint as to why attention is focused on certain particular phenomena as providing the basis from which the inductive step is taken. The scientist seldom simply amasses quantities of data, sifting through them hoping to find a regularity. Rather, he usually works in the context of some theory

---

[1] For a discussion of this problem see Ch. 4 of Hempel (1965). For a presentation of the view being reconstructed in the present chapter, see e.g. Ayer (1936).

which, as will be discussed later in this study, is not itself a regularity of the same sort as that being sought.

In various forms the problem of *confirmation* has received a great deal of attention in empiricist writings. The heart of this problem lies in the fact that the truth of the conclusion of a logical deduction does not imply the truth of the premises. In terms of the Deductive Model the problem is that the truth of the explanandum and statements of conditions, i.e. the truth of statements of the form $(Fa \wedge Ga)$, does not establish the truth of the law. And not only this, but since the law is conceived as an unrestricted universal statement, no one finite number of true statements of the above form provides any more support for it, or makes its truth any more probable, than does any other. But it may be pointed out that, if we do grant scientific laws as conforming to the model, then this problem of induction is not a problem for the empiricist, but for the scientist, for all that is demanded of the empiricist is that he provide a conception of science as it is actually practised.

But then it may be asked whether scientific laws do in fact have the form suggested by the Deductive Model. An examination of the nature of scientific laws, at least in the exact sciences, reveals that, rather than being expressed by statements having a truth-value, they are most often expressed as *equations* suggesting a numerical relationship among the values of certain parameters. And, where on the Deductive Model it is difficult to see how a statement discovered to be false might nevertheless continue to function as the expression of a law of nature, in science we find that laws expressed by equations are often retained even when it is realised that they have only a limited range of application.

While a positive account of the nature of scientific laws which is in keeping with the above observations will be given later, for present purposes it suffices to point out that the empiricist conception of science may be seen as being conceptually based on the Deductive Model, and that in this way it thus begins with a conception of scientific laws, rather than theories. In the next chapter the basis of the Popperian conception of science will be treated, and it too will be found to rest on the Deductive Model.

CHAPTER 3

THE BASIS OF THE POPPERIAN CONCEPTION OF SCIENCE

1. FALSIFIABILITY

The considerations of the previous chapter indicate that the logical positivist and logical empiricist views can be seen as attempting to demarcate (meaningful) science from (meaningless) non-science on the basis of verifiability and confirmability respectively. Popper's demarcation between science and non-science, on the other hand, is on the basis of falsifiability. For Popper, if there is no conceivable way that a statement can be shown to be false, while it might still be considered meaningful, it is not scientific but 'metaphysical.'

Seen most simply, on the empiricist conception the confirmation of scientific laws consists in the verification of observation statements entailed by them. On Popper's view, in its simplest form, laws or theories may be falsified via the determination of the truth of observation statements that contradict them. Thus where we can represent the empiricist conception by $A \vdash B$, where A is to include a general law, and B observational evidence, Popper's conception can here be represented by the formally equivalent:

(6) $\neg B \vdash \neg A.$

Here we see that the Popperian view lays stress on the idea that, while no amount of true observation statements of the sort B could verify the universal statement in A, the truth of one observation statement $\neg B$ should suffice to falsify A.[1]

The fact that at this primitive stage the schematisation of Popper's view is formally equivalent to that of the logical empiricist conception is worthy of note, for it suggests that the difference between the bases of the two views is more one of emphasis than of substance.[2]

---

[1] On this point, see Feyerabend (1974), p. 499.
[2] Cf. a similar remark by Carnap cited in Popper (1962), p. 254n.

Where the empiricist directs himself to the problem of what justifies our believing certain general claims of science to be valid, Popper points to a criterion – capable of being formulated within the empiricist conception – which should suffice to show them to be invalid. Of course the refutability of general claims in science was generally recognised before Popper made falsifiability his criterion of demarcation,[3] and, as will be seen below, Popper's main contribution beyond his demarcation criterion is his attempt to develop this idea in terms of laws (and theories) conceived of as general statements, i.e. in terms that can be represented by the Deductive Model.

The basis of the Popperian view as outlined to this point, in emphasising the falsifiability of general or universal claims, can be seen to have two serious shortcomings in comparison with a similarly simple presentation of the empiricist view. It affords a conception neither of the discovery of new laws (context of discovery), nor of the support of claims that certain laws exist (context of justification).[4] If it is viewed as providing a conception of discovery, such discovery is the discovery of mistakes; and granting that it affords a conception of justification (in a broad sense), such justification is the justification one might have in saying that something is wrong.

2. BASIC STATEMENTS AND BACKGROUND KNOWLEDGE

A first step in rendering Popper's conception more sophisticated parallels a move made by the empiricists in their development of the confirmability criterion of meaningfulness. It is the recognition that – in keeping with the conception of laws suggested by the Deductive Model – a universal statement entails an observation statement only when the former is conjoined with certain statements of initial conditions. In the case of the empiricist view this may be schematised by: $(L \wedge C) \vdash E$. In Popper's case the formulation is again equivalent,

---

[3] Cf. e.g. Poincaré (1902), pp. 150ff., Duhem (1906), pp. 180ff., and Campbell (1920), pp. 109 and 131.

[4] Lakatos would almost certainly have disagreed with this. On p. 375 of his (1968) he claims Popper to have focused attention on the problem of the discovery of hypotheses, when he has in fact focused attention on their refutation. But Lakatos' conception of the 'logic of discovery' is rather unusual – he sees it as the discipline of the rational *appraisal* of theories: in this regard see his (1970), p. 115.

but in keeping with the $\neg B \vdash \neg A$ schematisation in (6), it might first be presented as follows:

(7) $\quad \neg E \vdash \neg (L \wedge C)$.

This formulation of the basis of Popper's philosophy of science in terms of the Deductive Model makes it clear that his notion of falsification is not so straightforward as one might have hoped. Here, where we begin with the 'basic statement' $\neg E$,[5] we find that it does not entail the negation of the law or theory L, but rather entails the negation of the conjunction of L and the statement(s) of conditions C. Thus the determination of the truth of $\neg E$ would not suffice to falsify L.[6]

In order to obtain a situation in which L *is* falsified, Popper employs a line of thought that can best be represented by:

(8) $\quad (C \wedge \neg E) \vdash \neg L$.

In this formulation, which is still formally equivalent to the basis of the empiricist conception – i.e. to the Deductive Model – $(C \wedge \neg E)$, or, in the predicate calculus $(Fa \wedge \neg Ga)$, is the 'falsifying basic statement' deductively subsuming the negation of the law.[7] But it is obvious that this does not avoid the problem, for it is still the case that, just as the empiricist conception requires the truth (or confirmation) of C in order to confirm L, Popper requires the truth of C in order to falsify L.

Now, where the empiricists might want to say that statements of the sort C, e.g. 'This is copper being heated,' are capable of being observationally verified (or at least confirmed), Popper, in a move away from logical empiricism (logical positivism), argues that since such statements contain universal notions such as 'copper,' which themselves are based on certain theoretical presuppositions, they *cannot* be verified.[8] He suggests, in fact, that like general laws themselves, such statements can be falsified, but can be neither confirmed

---

[5] Cf. Popper (1959), p. 85n.
[6] For a similar point, see Duhem (1906), p. 185.
[7] Cf. e.g. Popper (1934), pp. 102 and 127, and (1959), p. 85n.
[8] Popper (1934), pp. 94–95; (1959), pp. 423–424&n. It may be noted that Carnap also adopts this stance in his (1936), pp. 425ff. In this regard cf. also Campbell (1920), p. 43.

nor in any way established as true. Thus for Popper, in a situation such as that depicted by (8), statements of the sort C are to be 'background knowledge,' which for the purpose of falsifying L are to be *tentatively* accepted as unproblematic.[9] But if we ask why they (or the basic statements containing them) are to be so accepted, we are told that in certain cases it is because they have survived similar attempts at falsification.[10] On Popper's reasoning, however, this leads to an infinite regress, which is ultimately ended only by the consensus of scientists. But the basing of falsifications on consensus, while perhaps a fair procedure, is hardly an objective one, and does not suffice to distinguish science from other pursuits.

In any case, the fact that Popper suggests that statements C cannot be established as true, for whatever reason, means that general statements cannot be established as false.[11] And this in turn means that Popper cannot claim to have provided a criterion for distinguishing science from non-science on the basis of the falsifiability of the claims made by the former.[12]

3. CORROBORATION, SEVERITY OF TESTS, AND THE FALSIFIABILITY OF THE EMPIRICAL BASIS

One of the advantages Popper claims for his view is that it solves the problem of induction. As was mentioned in the previous chapter, the notion of induction may be seen to have two distinct applications: one to the discovery of new laws, and one to the justification of claims that a particular law has in fact been found. Popper focuses on the latter application, and believes himself to have solved the problem of induction through suggesting that our reason for accepting certain theories in science is not because they have been established as true, but because they have survived (severe) attempts at falsification.[13]

---

[9] Cf. e.g. Popper (1962), pp. 390f.

[10] Popper (1934), p. 104.

[11] For a similar criticism see Deutscher (1968), p. 280; see also Duhem (1906), p. 185.

[12] Note that the present criticism does not preclude some sort of testability criterion of demarcation, but relates directly to Popper's *falsifiability* criterion as it is to be understood in the context of his conception of scientific laws.

[13] Cf. e.g. Popper (1973), p. 8.

And theories that have survived such tests are, as a result, to be considered 'corroborated.'

Of course few people, if any, would claim that the determination of true consequences of an empirical law (conceived of as a general statement) would *establish* the law as true; and, as it has been presented in the previous chapter, the empiricist problem of confirmation concerns rather the formulation of a conception in which, roughly speaking, the determination of the truth of consequences of the law would add support to the claim that the law is true.[14] Nevertheless, Popper for the most part directs his critical arguments in this context against whomever it might be that believes that such laws can be completely verified.[15]

However, at one point Popper does treat of the case where the truth of the consequences of a law may be thought to add confirmatory support to the law. His argument here can be clearly presented in terms of the Deductive Model. If we assume a law to have the form $\forall x \, (Fx \rightarrow Gx)$, then for a particular object a, the law implies that $(Fa \rightarrow Ga)$. Statements having the latter form Popper calls 'instantial statements,' and suggests that since they, or their logical equivalents ($\neg Fa \vee Ga$) and $\neg (Fa \wedge \neg Ga)$, are verified wherever $\neg Fa$ is the case – i.e. almost everywhere – "these instantial statements *cannot play the role of test statements* (or of potential falsifiers) which is precisely the role which basic statements are supposed to play."[16]

Popper is perhaps here overlooking the fact that these 'instantial statements' are precisely the statements which must be determined true in order to falsify basic statements; and his argument that such statements cannot function as test statements in the context of general laws applies just as well in this other context, and results in his implicitly denying scientific status to his own basic statements. But, in any case, if we are to interpret Popper's argument as a criticism of the empiricist notion of confirmation we see that it is wide of the mark, for, as has been pointed out in the previous chapter, on the empiricist view it is not statements of the form $(Fa \rightarrow Ga)$ that

[14] Cf. Hempel (1966), pp. 116f.

[15] Popper's only reference in this regard is to an article written by Gilbert Ryle in 1937: cf. Popper (1973), p. 9.

[16] Popper (1959), p. 101n.

should be thought to confirm a law, but statements of the form $(Fa \wedge Ga)$.[17] And the verification of such statements is no more problematic than is the verification of Popper's basic statements.

But the problem of confirmation, as described in the previous chapter, still remains, and so we might investigate whether Popper has succeeded in solving or avoiding it via the employment of his own notion of corroboration. What is being required of him here then is the presentation of a conception not involving the notion of induction, in terms of which we can understand how the passing of (severe) tests should make more reasonable the acceptance of a particular scientific law.

In keeping with Popper,[18] his definition of corroboration[19] may also be presented in terms of the Deductive Model:

$$(9) \qquad C(L, E, C) = \frac{P(E, LC) - P(E, C)}{P(E, LC) - P(LE, C) + P(E, C)}.$$

Here "$C(L, E, C)$" stands for: 'the degree of corroboration of law L by explananda E, in the presence of conditions (background knowledge) C'; and, for example, "$P(E, LC)$" means: 'the relative probability of the explananda E, given the truth of the conjunction of L and C.'

Disallowing induction, the above suggests that the degree of corroboration of all laws would be the same: $1-0/1-0+0 = 1$. (This is not surprising considering that Popper sees no possibility of attaching numerical values other than 0 and 1 to his measures of probability).[20] But what this means is that Popper has not succeeded in providing a metric for degree of corroboration.[21]

However, all that is being asked of Popper here is that he provide a non-inductive conception of how the passing of tests should corroborate a theory, and so his failure to provide a metric may be overlooked, and we might investigate instead the ideas lying behind his

---

[17] We note that it is normally statements of this latter form which are called instantial, and that what Popper calls instantial statements are actually hypothetical statements.

[18] Cf. Popper (1962), pp. 390f.

[19] Popper (1959), p. 400n.

[20] Cf. Popper (1962), p. 397.

[21] In this regard cf. Lakatos (1968), p. 396.

definition, to see if such a conception may be found there. Thus we note that if it were not for the middle term in the divisor (which Popper especially inserts ostensibly in order to satisfy certain intuitive desiderata), the degree of corroboration of L would depend solely on the value of $P(E, C)$;[22] i.e., as the probability of E, given background knowledge C, rises, so should the degree of corroboration of L.

Underlying Popper's thinking here is that as each test of L is passed, the results of the test are to be added to the background knowledge; thus, the more tests L passes, the greater is the probability that it will pass the next one. In this way too, on Popper's view, the *severity* of each successive test is to decline. In Popper's own words:

[I]f a theory stands up to many such tests, then, owing to the incorporation of the results of our tests into our background knowledge, there may be, after a time, no places left where (in the light of our new background knowledge) counter examples can with a high probability be expected to occur. But this means that the degree of severity of our test declines.[23]

And, as the severity of tests declines and the probability of E given C rises, so too rises the degree of corroboration of L.

However, while this line of thinking seems intuitively reasonable, it can be of no use to Popper, since it is inductive, and, consequently, so are Popper's concepts of corroboration and severity of tests. Not only this, but if we stop to consider the way in which Popper conceives of a theory being tested we see that his notion of corroboration, or the passing of tests, is essentially identical to the empiricist notion of confirmation. For the empiricist the determination of the truth of a statement of the form $(Fa \wedge Ga)$ should provide confirmatory support for the truth of a law of the form $\forall x\,(Fx \rightarrow Gx)$. For Popper, such laws are tested by attempting to discover acceptable falsifying basic statements of the form $(Fa \wedge \neg Ga)$. Both notions require the determination of the truth of Fa, and if it is also found

---

[22] On this point cf. ibid., p. 410n.

[23] Popper (1962), p. 240. Popper's notion of severity of tests, given here, ought not be confused with his notion of testability, which is essentially the same as his concept of content – though Popper himself sometimes conflates these two notions: cf. e.g. Popper (1959), p. 374.

that $\neg Ga$ is the case, then the law is falsified. But if its negation, Ga, is the case then, on the empiricist view, the law has been confirmed, and, on Popper's view, it has passed its test and has consequently been corroborated. As a result of the above considerations then we see that Popper is not at all warranted in claiming to have solved the problem of induction, even when it is limited to the context of justification. His notions of corroboration and severity of tests are both inductive, and do not differ in any significant way from the logical empiricist conception of (degree of) confirmation; and his conception of the testing of a theory is also essentially the same as that of the empiricists.

But if we question whether scientific laws actually do have the form suggested by the Deductive Model, as has been done at the end of the previous chapter, then questions of the sort treated in this chapter, framed in terms of the model, become mainly of academic interest. That this is so is even more evident when it is realised that Popper makes no distinction between scientific laws and theories, taking both to have the form suggested by the model.[24]

It is hoped that the efforts of the last two chapters have succeeded in showing both Popper's conception and that of the empiricists to be formally based on the Deductive Model. Both views take scientific laws to be universal or general statements having a truth-value, and the extent to which we know this truth-value is to be based on our knowledge of the truth-values of certain explananda and statements of conditions. Given true statements of conditions (assuming their truth determinable) and a false explanandum, the law is considered false, whereas true statements of conditions and true explananda are to confirm or corroborate the law. And the essential difference between the empiricist and Popperian views as developed thus far may be seen to lie in the empiricists' normally applying the model to cases where the explanandum may be considered true, where Popper usually applies it to cases in which the explanandum should be false. In the next two chapters this difference in the application of the model will be seen to lead to strikingly divergent accounts of the nature of scientific progress.

---

[24] See e.g. Popper (1934), Ch. 3, where he treats theories in this way, explicitly stating on p. 59 that "Scientific theories are universal statements." In this regard see also Popper (1959), pp. 426ff.

CHAPTER 4

THE LOGICAL EMPIRICIST CONCEPTION
OF SCIENTIFIC PROGRESS

1. A FORMAL CRITERION OF PROGRESS

As mentioned in Chapter 1, both Popper and the empiricists advocate the use of the Deductive Model as a model of the explanation of particular occurrences. But the empiricists go one step further and suggest its employment as a model of the explanation of laws by higher-level theories. For example, Morris Cohen and Ernest Nagel have said:

Scientific explanation consists in subsuming under some rule or law which expresses an invariant character of a group of events, the particular event it is said to explain. Laws themselves may be explained, and in the same manner, by showing that they are consequences of more comprehensive theories.[1]

In presenting the basis of the logical empiricist conception of science it was shown how the Deductive Model, originally introduced as a model of explanation, can also function as a model of induction and confirmation in the contexts of discovery and justification respectively. Here it may be put to the same use. Thus, on the empiricist conception, scientific advance may be seen to consist in the discovery of higher-level laws or theories – conceived as universal statements – which deductively entail lower-level ones.[2] Following Cohen and Nagel this means that here the place of the explanandum in the Deductive Model is to be filled, not by an empirically verifiable statement, but rather by a general law. This law L is to be

---

[1] Cohen & Nagel (1934), p. 397. But cf. Campbell (1921), p. 80: "To say that all gases expand when heated is not to explain why hydrogen expands when heated; it merely leads us to ask immediately why all gases expand. An explanation which leads immediately to another question of the same kind is no explanation at all."

[2] In his concentration on the work of Carnap, Lakatos misses this empiricist conception of the growth of knowledge. See Lakatos (1968), p. 326.

derivable from some theory L_1 in conjunction with particular statements of conditions C_1, which can also have the character of universal statements. The inductive discovery and confirmatory justification of L_1 is thus partly to depend on that of L. This may be schematised as below:

(10) $(L_1 \wedge C_1) \vdash L.$

To make explicit the deductive relation from L_1 and C_1 to L, (10) may be presented directly in the predicate calculus thus:

(11) $\forall y\, (Hy \rightarrow Gy) \wedge \forall z\, (Fz \rightarrow Hz) \vdash \forall x\, (Fx \rightarrow Gx).$

And (10) and (11) may both be applied to an extension of the example used earlier:

(12)	L_1	$\forall y\, (Hy \rightarrow Gy)$	All metals expand when heated
	C_1	$\forall z\, (Fz \rightarrow Hz)$	Heated copper is a metal
thus,	L	$\forall x\, (Fx \rightarrow Gx)$	All copper expands when heated.

If we conjoin to (10) the original schematisation of the Deductive Model we get:

(13) $(L_1 \wedge C_1) \vdash L$ and $(L \wedge C) \vdash E.$

And (13) itself implies the original form of the model (as expressed in the sentential calculus):

(14) $L_1 \wedge (C \wedge C_1) \vdash E.$

In order that L_1 be more than a formal embellishment – that its discovery should constitute actual progress – a further requirement may be set. It is that L_1 have confirming instances, or make predictions, beyond those of L. This requirement can be easily handled in terms of the Deductive Model by conjoining to (13):

(15) $(L_1 \wedge C_2) \vdash L_2$ and $(L_2 \wedge C_3) \vdash E_1,$

where, as throughout, difference of subscript is to suggest difference of referent. And the conjunction of (13) and (15) may be exemplified by:

(16) L_1 All metals expand when heated

C_1 Heated copper is a metal		C_2 Heated tin is a metal

thus, L All copper expands   thus, L_2 All tin expands
          when heated   when heated

C This is copper being C_3 This is tin being
    heated   heated

thus, E This (copper) expands   thus, E_1 This (tin) expands.

The schematisation of (16) in turn implies:

(17) $L_1 \wedge (C \wedge C_1 \wedge C_2 \wedge C_3) \vdash E \wedge E_1$,

which, except for there being more than one explanandum, is also of the form of the Deductive Model as expressed in the sentential calculus.

On the basis of the above a logical empiricist criterion of scientific progress can be formulated in terms of the Deductive Model. Provided that all of the statements concerned are distinct and have been verified or confirmed, we should say that one law or theory L_1 is a progression beyond another, L, only if (a) L_1, in conjunction with certain statements of conditions C_1, entails L, and (b) L_1, in conjunction with statements of conditions C_2, entails other verified or confirmed statements which L alone in conjunction with C_2 does not entail, e.g., L_2.

Thus, in keeping with this criterion, we read in Hempel:

[T]he uniformity expressed by Galileo's law for free fall can be explained by deduction from the general laws of mechanics and Newton's law of gravitation, in conjunction with statements specifying the mass and radius of the earth. Similarly, the uniformities expressed by the law of general optics can be explained by deductive subsumption under the principles of the wave theory of light.[3]

---

[3] Hempel (1962), pp. 100–101; see also Hempel (1965), pp. 343ff. Though Hempel intends here to be providing examples of how the Deductive Model can be applied to the case of theories explaining laws, in a footnote he goes on to concede that: "strictly speaking, the theory (Newton's) contradicts Galileo's law, but shows the latter to hold true in very close approximation within a certain range of application. A similar relation obtains between the principles of wave optics and those of geometrical optics." (1962), p. 101n. In regard to this point see the concluding paragraph of the present chapter and p. 27 below.

2. THE PROBLEMS OF THEORETICAL TERMS
AND CORRESPONDENCE RULES

As mentioned in Chapter 2, on the logical positivist conception of
meaningfulness a (non-tautological) statement is meaningful only if
it is verifiable by direct experience. Thus, for the positivist, meaningful claims should involve no 'descriptive' terms, i.e. no non-logical
terms, the referents of which are not directly observable. While the
empiricist conception, on the other hand, replaces the verifiability
requirement with that of confirmability, confirmable statements
(empirical laws) should still be such as to admit no non-logical terms
having unobservable referents. But when it is realised that scientific
theories contain terms such as "electron," the referents of which are
not directly observable, the problem arises as to how statements containing such terms can be construed as meaningful, granting that the
empiricist programme wants to afford this status to all scientific
terms. This is the problem of theoretical terms,[4] which also arises in
the case of mensural terms such as "temperature," and in the case of
terms such as "magnet," the referents of which have dispositional
properties.

Where the empiricist problem of theoretical terms concerns the
transfer of meaning from the empirical level to the theoretical, their
closely allied problem of correspondence rules may be said to start at
the theoretical level, and to concern the question of how theoretical
notions are to apply to empirical situations. On the present reconstruction the problem of correspondence rules may thus be seen to be
that of showing how scientific theories, taken to contain theoretical
terms as essential elements, can logically entail empirical laws,
which are to contain no theoretical terms.[5]

---

[4] Thus we see that it is the existence of theoretical terms which is the problem; the
distinction between observational and theoretical terms was not made in order to
solve the problem, as has been suggested e.g. in Putnam (1962) and Achinstein
(1965). In this regard see pp. 131–133 below.

[5] Cf. Carnap (1966a), p. 232: "The statement that empirical laws are derived from
theoretical laws is an oversimplification. It is not possible to derive them directly
because a theoretical law contains theoretical terms, whereas an empirical law contains only observable terms. This prevents any direct deduction of an empirical law
from a theoretical one."

One approach to the solution of problems of this sort has been to suggest means of eliminating theoretical terms in such a way as to retain the empirical consequences of the theory, as conceived on the empiricist view. For example, following a suggestion of F. P. Ramsey,[6] if we consider a scientific theory in conjunction with a statement of conditions to have the form:

(18) $\forall x\,(Tx \rightarrow Ox) \wedge Ta,$

where "T" is a theoretical term, "O" an observational term, and a some entity which it can be supposed has the property represented by the theoretical term, then "T" can be eliminated by reformulating (18) in terms of one of its consequences in the second-order predicate calculus:

(19) $\exists \phi\,(\forall x\,(\phi x \rightarrow Ox) \wedge \phi a).$

Here the predicate constant T has been dropped in favour of a second-order (bound) predicate variable ϕ. And we see that (19) has in fact retained the empirical consequences of (18), namely, Oa.

However, while the 'Ramsey-sentence' (19) has succeeded in preserving the empirical consequences of (18) without employing the theoretical term "T," its doing so has required the introduction of the term "ϕ." And, as Hempel has said:

[T]his means that the Ramsey-sentence associated with an interpreted theory T' avoids reference to hypothetical entities only in letter – replacing Latin constants by Greek variables – rather than in spirit. For it still asserts the existence of certain entities of the kind postulated by T', without guaranteeing any more than does T' that those entities are observables or at least fully characterizable in terms of observables. Hence, Ramsey-sentences provide no satisfactory way of avoiding theoretical concepts.[7]

But once again, more important than such questions as whether the 'Ramsey method' can be employed so as to overcome the empiricists' problems of theoretical terms or correspondence rules is the

---

[6] Cf. Ramsey (1931). We note however that Ramsey does not himself suggest the procedure given here as a means of eliminating theoretical terms, but as "[t]he best way to write our theory" (p. 231) – i.e. as the best way to make more salient those aspects of his example which he considers important.

[7] Hempel (1965), p. 216.

fact that the 'theories' to which the method has been applied (by Ramsey himself, for example) bear little resemblance to actual scientific theories. And until such methods are shown to have application to real theories, even positive results would mean little with regard to our understanding of the nature of science.

3. THE PROBLEMS OF MEANING VARIANCE AND CONSISTENCY

If we construe the empiricist notion of scientific progress as involving the relation between theories, rather than between a theory and a law, problems arise which have a more direct bearing on the topic of the present study. As has been pointed out by Paul Feyerabend, the conception of scientific advance as consisting in the deductive subsumption of earlier theories by later ones presupposes that the meanings of the terms common to the theories remain constant.[8] If this were not so, then such derivations (granting them possible) would at most tell us something about the syntactical relations between the theories viewed simply as parts of one and the same abstract system – they would tell us nothing about the semantical relations between them, i.e., about how the theories relate to each other as entities providing information about some aspect of the real world. Thus what will here be called 'the problem of meaning variance' is that in certain important cases of theory succession in actual science terms do undergo a change in their meanings (e.g. Newtonian "mass" vs. Einsteinian "mass"); and the conception of scientific theories as being related in ways suggested by the Deductive Model – which provides only the purported linguistic form of theories – is unable to take account of this fact.

A second point, stated clearly by Duhem, and later emphasised by Popper, and, following him, by Feyerabend,[9] is that in attempting to provide accounts of the same realm of phenomena it is often the case that successive theories (or laws) are inconsistent with one another. The above authors have expressed this by saying that such theories contradict each other. Later in this study it will be suggested that

---

[8] See e.g. Feyerabend (1963), pp. 16ff.
[9] See Duhem (1906), p. 193; Popper (1949), pp. 357–359, and (1957a), pp. 197ff; and e.g. Feyerabend (1963), pp. 20ff.

theory conflict does not take the form implied by the term "contradiction;" but here conceiving of science in the context of the Deductive Model, we see that in the case where two successive theories do conflict, unless the succeeding theory (in conjunction with the relevant statements of conditions) is self-contradictory, it cannot formally entail its predecessor. This 'problem of consistency,' i.e. this inability to provide an account of theory conflict, places severe limitations on the empiricist conception of scientific progress. In the next chapter Popper's conception of scientific progress will be considered, and will be seen to involve the Deductive Model in such a way as to afford a conception of theory conflict.

CHAPTER 5

THE POPPERIAN CONCEPTION
OF SCIENTIFIC PROGRESS

1. CONTRADICTION

In Chapter 3 the basis of Popper's philosophy of science, including his notions of basic statement, background knowledge, and corroboration, were shown to rest on the Deductive Model, in which scientific laws are conceived to be universal statements of the form: $\forall x\,(Fx \rightarrow Gx)$. And in Chapter 4 a logical empiricist conception of progress was presented, also in terms of the model. In the present chapter it will be shown that Popper's attempts to provide a conception of progress likewise rely on the Deductive Model, the basic difference between his view and that of the empiricists being that, where the latter see succeeding theories as logically entailing their predecessors, Popper sees such theories as contradicting one another. Thus, for example, Popper claims that:

[F]rom a logical point of view, Newton's theory, strictly speaking, contradicts both Galileo's and Kepler's .... For this reason it is impossible to derive Newton's theory from either Galileo's or Kepler's or both, whether by deduction or induction. For neither a deductive nor an inductive inference can ever proceed from consistent premises to a conclusion that formally contradicts the premises from which we started.[1]

In its simplest form, Popper's conception of the relation between successive theories can be schematised in terms of the Deductive Model by:

(20) $(L \wedge C) \vdash E$ and $(L_1 \wedge C) \vdash \neg E,$

where L and L_1 are the 'theories' in question, C are statements of conditions, E is an explanandum statement, and $\neg E$ its negation. (It may be noted that while on the empiricist conception theories can be distinguished from laws in that they have the latter as their explananda,

---

[1] Popper (1957a), p. 198; see also Popper (1975), pp. 82–83&n.

as mentioned in Chapter 3 on Popper's conception no distinction is drawn between laws and theories. In what follows the term "theory" will be used in the sense intended by Popper.)

The nature of Popper's conception of theory conflict can be further clarified by deriving from (20):

(21) $(L \land C) \vdash \neg (L_1 \land C)$.

Here, however, it becomes clear that in order to obtain a contradiction between L and L_1 Popper requires the truth of statements C, just as he did in the case of his falsifiability criterion.[2] And so, "strictly speaking," L and L_1 do *not* contradict one another. But in the present chapter this problem will be set aside, i.e. it will be allowed that C can in some way be determined true, and that Popper has thus succeeded in providing a notion of theory conflict (as contradiction) where the empiricists have not.

Thus we see that where the essential relation between theories on the empiricist view is that of deductive subsumption, on Popper's view it is formal contradiction. The empiricists succeed in providing a conception of how one theory might be considered scientifically superior to another – but their view fails to account for theory conflict. Popper has provided a conception of theory conflict – now his is the task of showing how one of two such conflicting theories is to be conceived as constituting an advance over the other.

It might be thought that Popper should here simply say that, of two theories, the superior is the one which has not been falsified (or 'refuted'). But this will not do, for it does not treat of the interesting cases in which both theories are false. Were Popper to lay all his stress on falsification as a criterion for distinguishing good theories from bad, then all refuted theories would be equally unacceptable; and since on Popper's account the history of science shows that theories are constantly being falsified, he would have to say that science is not progressing at all, and thus commit himself to scepticism. Consequently, Popper must provide criteria allowing us to determine, at least in principle, which of two contradicting false theories is a progression beyond the other.

---

[2] Formally, this is due to the conception of theories as hypothetical statements. Thus, for example, $\forall x (Fx \rightarrow (Gx \land \neg Gx))$ is contradictory only if Fx is assumed true for some x.

In his attempt to provide such criteria there are two main factors that Popper must take into account. One has to do with how close each of the false theories is to being true. But this is not alone a sufficient basis for a criterion of progress, for it implies that some relatively trivial generalisation might be a progression beyond some comprehensive, albeit refuted, theory. Consequently a second factor must be considered, namely the comprehensiveness or non-triviality of a theory. This latter factor constitutes Popper's concept of *content*, and a combination of both factors gives his notion of *verisimilitude*. Below, each of these concepts will be discussed in detail.

2. CONTENT

In Chapter 10 of *Conjectures and Refutations* Popper says:

My study of the *content* of a theory (or of any statement [*sic*] whatsoever) was based on the simple and obvious idea that the informative content of the *conjunction*, *ab*, of any two statements, *a*, and *b*, will always be greater than, or at least equal to, that of any of its components.

Let *a* be the statement 'It will rain on Friday'; *b* the statement 'It will be fine on Saturday'; and *ab* the statement 'It will rain on Friday and it will be fine on Saturday': it is then obvious that the informative content of this last statement, the conjunction *ab*, will exceed that of its component *a* and also that of its component *b*.[3]

On the logical empiricist conception of progress the question of the relative content of theories is easily handled: the higher-level theory has a greater content than the theory or law it subsumes. But Popper, in seeing theories as contradicting, requires some other means of determining relative content. It is peculiar then that the idea upon which his concept of content is based is formulated along empiricist lines rather than along his own. The conjunction *ab* deductively subsumes each of *a* and *b* and thus has a content greater than or equal to either of them – but this is of no help in conceiving how the content of one proposition might be greater than that of another when the two of them contradict.

Nevertheless, from this basis Popper goes on to explicate his notions of *probability of a theory*, and *testability*. He suggests that if

---

[3] Popper (1962), pp. 217–218.

a and *b* are empirical propositions, the probability that both are true is less than the probability that either alone is true. And since the falsification of either conjunct will falsify the whole, he says of *ab* that it is more testable than either *a* or *b* alone.[4] Popper considers this to provide a criterion of potential satisfactoriness of theories, and says:

The thesis that the criterion here proposed actually dominates the progress of science can easily be illustrated with the help of historical examples. The theories of Kepler and Galileo were unified and superseded by Newton's logically stronger and better testable theory, and similarly Fresnel's and Faraday's by Maxwell's. Newton's theory, and Maxwell's, in their turn, were unified and superseded by Einstein's.

Needless to say, this picture is not at all in keeping with Popper's view of successive theories as contradicting. It is more an empiricist conception than a Popperian one, and its acceptance would immediately raise the question of how to account for theory conflict.

The problem of conceiving of increase in content in terms either of conjunction or deductive subsumption, while at the same time conceiving of the relation between competing theories as being that of contradiction, runs throughout Popper's writings on the topic of content. Though he cannot maintain both conceptions, and the latter is central to his philosophy of science, his explications of how one theory might have more content than another are almost always made along the lines of the former.

But in elaborating the concept of content Popper does provide a conception via which the sizes of the respective contents of contradicting theories might possibly be determined. This elaboration consists in a distinction between what Popper calls 'logical content' and 'empirical content': "the *logical content* of a statement or a theory *a* is the class of all statements which follow logically from *a*, while ... the *empirical content* of *a* [is] the class of all basic statements that contradict *a*."[5]

Popper's distinction between logical content and empirical content is essentially the distinction between the empiricists' notion of a verifiable statement E derivable from a law L in conjunction with

---

[4] Ibid., pp. 218ff. Note that, as mentioned in Chapter 3, Popper's notion of testability differs from his notion of severity of test. Quote following, ibid., p. 220.

[5] Ibid., p. 232.

statements of conditions C, and Popper's own notion of a (falsifying) basic statement $(C \wedge \neg E)$ which contradicts L. The logical content of a theory is more its actual content – what the theory (plus statements of conditions) contains, and it is partly in terms of this notion that Popper later defines verisimilitude. At this point however, since there is no observational statement E that follows from L alone, Popper feels justified in claiming that the empirical content of L consists rather in the class of basic statements contradicting L.[6] But, as was seen above, Popper needs statements of conditions C for contradiction equally as much as the empiricists need them for derivation, so the fact that E does not follow from L without C does not warrant granting the status of empirical content to the class of basic statements and not to the class of verifiable statements. (This conclusion is further supported by the fact that, seeing as on Popper's account competing theories contradict, at least part of the logical content of each must be in the empirical content of the other.)

While Popper denies the status of empirical content to the set of statements derivable from a theory, he goes on to argue that although the class of basic statements contradicting a theory is actually excluded by the theory, it may nevertheless be called the (empirical) *content* of the theory for the reason that its measure varies with that of the set of derivable statements:

That the name 'empirical content' is justifiably applied to this class is seen from the fact that whenever the measures of the empirical contents, $ECt(t_1)$ and $ECt(t_2)$, of two empirical (i.e. non-metaphysical) theories, t_1 and t_2, are so related that

(1) $ECt(t_1) \leq ECt(t_2)$

holds, the measures of their logical contents will also be so related that

(2) $Ct(t_1) \leq Ct(t_2)$

will hold; and similar relations will hold for the equality of contents.

Thus through a measure of the respective classes of basic statements of two theories we should be able to determine the relative sizes of their (logical) contents. This process need not involve the deductive subsumption or conjunction of theories and so can allow for their

---

[6] Ibid., p. 385; quote following, ibid.

contradicting; what it requires instead is some other means of determining which theory excludes the more basic statements.

But how, even in principle, is this determination to be made? Popper does not begin to answer this question, and in fact says that he sees no possibility of attaching numerical values other than 0 and 1 to measures of content.[7] Thus, at least as far as Popper is concerned, there is no measure of the classes of basic statements of each of two theories that can tell us which theory has the greater content.

Content as Varying Inversely to Absolute Probability

After presenting the concept of content in terms of conjunction (Chapter 10, *Conjectures and Refutations*), Popper goes on to say that content varies inversely to probability:

Writing $Ct(a)$ for 'the content of the statement a,' and $Ct(ab)$ for 'the content of the conjunction a and b,' we have

(1) $Ct\,(a) \leq Ct\,(ab) \geq Ct\,(b)$.

This contrasts with the corresponding law of the calculus of probability,

(2) $p\,(a) \geq p\,(ab) \leq p\,(b)$,

where the inequality signs of (1) are inverted. Together these two laws, (1) and (2), state that with increasing content, probability decreases, and *vice versa*; or in other words, that content increases with increasing *im*probability.[8]

As mentioned above, Popper's line of thinking here is that the probability of the conjunction of two empirical propositions being true is less than the probability of either alone being true. Implicit in this is that the truth-value of neither of the propositions is known – in other words, that the probability being considered is an a priori, or absolute, probability. Now, granting that content might vary inversely to absolute probability as Popper suggests, if a means of determining the respective absolute probabilities of two contradicting theories could be provided, then this would in turn afford a means of judging the relative sizes of their contents.

---

[7] Ibid., p. 397.
[8] Ibid., p. 218.

Essentially the same notion of absolute or a priori probability as considered by Popper is employed by a number of philosophers writing in the spirit of the logic of science, including Rudolf Carnap and William Kneale.[9] These philosophers would agree, following J. M. Keynes, that the a priori probability of each of the following formulae is greater than, or at least equal to, that of the one(s) above it:

(22) $\forall x\ (Fx \rightarrow (Gx \wedge Hx))$
        $\forall x\ (Fx \rightarrow Gx)$
        $\forall x\ ((Fx \wedge Ix) \rightarrow Gx).$

This is so because each formula is implied by those preceding it. Keynes warns, however, that we may not be able to compare the respective a priori probabilities of two generalisations unless the antecedent of the first is included in that of the second, and the consequent of the second is included in that of the first, as above.[10] This latter condition of course requires that, if the generalisations concerned are each consistent, then they do not contradict one another.

As a consequence of this, if Popper hopes to place any weight at all on the idea of one of two contradicting theories having a greater a priori probability than the other, then the onus is on him first to show how such a difference is to be conceived. But, as a matter of fact, he never even tries to show this. He instead takes the view that all laws or theories have absolute (and relative)[11] probability zero.[12] And so we must conclude that a measure of content is not to be found via consideration of Popper's notion of a priori probability.

Zero Probability and the Fine-Structure of Content

Popper's view that all theories have zero probability is arrived at in arguing against empiricists such as Carnap, who have claimed that theories may have a higher or lower probability depending on their

---

[9] Cf. e.g. Carnap (1966b), and Kneale (1964).

[10] Cf. Keynes (1921), p. 225.

[11] Cf. Popper (1959), p. 364. The notion of relative probability plays a small role in the present considerations. This is so because what is being demanded of Popper is a criterion for determining the relative superiority of false theories; and all such theories (on any account in which they are conceived as universal statements) must have relative or a posteriori probability zero.

[12] Cf. Popper (1958), p. 192; and (1959), pp. 363ff. and 373.

degree of confirmation. Popper's argument is in itself reasonable, and it is that any law or theory conceived as a universal statement entails an infinite number of singular statements each with a probability less than one, and that since the product of the probabilities of these statements will equal zero, the probability of the law itself must equal zero.[13] But this leaves him in the awkward position of having to explain how the size of the content of theories might differ, considering that it varies inversely to probability.[14] In an attempt to rescue his notion of content Popper thus claims that, though the probabilities of any two universal statements must equal zero, an analysis of the 'fine-structure' of their contents may allow us to determine which content is the greater.

In order to exemplify how the content of two laws or theories can differ in spite of their both having zero probability, Popper suggests that we consider as laws: 'All planets move in circles' and 'All planets move in ellipses.'[15] He then goes on to say that, since the former entails the latter, it has the greater content and, furthermore, the greater degree of testability, since any test of the entailed law is also a test of the entailing one, but not vice versa.

Here again we see the employment of the logical empiricist conception of succeeding theories deductively subsuming their predecessors, rather than contradicting them. Though Popper does not claim actual scientific theories to be related in exactly this way, he nevertheless suggests further that:

Similar relationships may hold between two theories, a_1 and a_2, even if a_1 does not logically entail a_2, but entails instead a theory to which a_2 is a very good approximation. (Thus a_1 may be Newton's dynamics and a_2 may be Kepler's laws which do not follow from Newton's theory, but merely 'follow with good approximation'; ....)[16]

Here we might expect Popper to conclude his argument by saying something to the effect that the actual case is sufficiently similar to that of deductive subsumption to warrant granting that a_1 has a greater content than a_2 (though Carnap could then employ the same

---

[13] Popper (1959), pp. 364ff.
[14] Lakatos too notes that theories' having zero probability means that the notion of probability cannot be used to determine their content: see Lakatos (1968), p. 379.
[15] Popper (1959), p. 373.
[16] This quote and the next, ibid., p. 374.

argument to suggest that a_2 has a greater a priori probability than a_1).
Or, considering his emphasis on the notion of testability in the present context, it might be thought that Popper should here say that
even in the actual case a_1 is better testable than a_2, and consequently
may be said to have the greater content. But what he does say is the
exact converse – immediately following the above quotation Popper
concludes that, "Here too, Newton's theory is better testable, because
its content is greater." But the question at issue does not directly
concern the concept of testability – it concerns the concept of content. We see here, however, that Popper does not claim a_1 to have a
greater content than a_2 on the basis of its being better testable, for, as
he has himself suggested earlier, this should imply that any test of a_2
is also a test of a_1. And if he were to apply this reasoning to the present case, saying that any test of Kepler's laws is also a test of
Newton's theory, he would be forced to admit that the falsification of
Kepler's laws would entail the falsification of Newton's theory. This,
of course, he cannot afford to do. Thus, aside from his not having provided a positive conception of difference of content in the case where
successive theories are seen to contradict one another, Popper's
attempt to overcome the problems raised by the zero probability of
theories via the employment of the concept of logical entailment
does not solve these problems, but only gives rise to further ones.

Nowhere in his dealings with the notion of content does Popper
provide a conception of how to determine which of two contradicting
theories has a greater content than the other. This is of major importance to his account of scientific progress, since according to it successive theories do contradict. What he does instead is present the
notion of content in terms of the concepts of deductive subsumption
and conjunction, which do not fit his view of science, and in terms of
such concepts as those of absolute probability and testability, which
are just as problematic as the notion of content itself.

3. VERISIMILITUDE

The second major factor that must be taken into account in Popper's
conception of scientific progress is that of 'nearness to the truth.' On
Popper's view, though two contradicting theories may both have

been refuted, one of them might nevertheless be closer to the truth than the other.

Two Conceptions of Nearness to the Truth

Taking theories to be entities that are either true or false, the idea of one false theory being closer to the truth than another may be thought of in one of two ways: either the theories may both be considered as being completely false, while one is closer to being true than the other; or they may be considered as each being partly true and partly false, while one is more true (or less false) than the other. In the former conception truth is not something attained, but rather something approached; in the latter conception truth, though attained, is not comprehensive.

Following Popper, these two conceptions may be exemplified by statements from ordinary language. An example of the first notion might go as follows: Given that the time at present is 12 o'clock, the statement 'The time is now 11 o'clock,' while completely false, is closer to the truth than the also false 'The time is now 10 o'clock.'[17] And the second view might be exemplified by 'It always rains on Saturdays' and 'It always rains on Sundays.'[18] In this case Popper would consider both statements to be partly true in that it has rained at least once on each of these days. The former statement might be said to be nearer to the truth if it has rained more often on Saturdays than on Sundays.

The criterion for determining the distance from the truth of each of the completely false statements is an empirical one, involving the measurement of time (in this way it is similar to the criterion of scientific progress to be suggested later in this study); and it is difficult to see how it might be captured employing the tools afforded by the Deductive Model. On the other hand, the criterion of nearness to the truth in the case where the statements are each conceived to be partly true appears, at least prima facie, to be susceptible of explication in

---

[17] Cf. Popper (1973), pp. 55ff.
[18] Cf. Popper (1962), p. 233.

terms of the model; and it is along this latter line that Popper develops his notion of verisimilitude.[19]

Verisimilitude and Logical Content

Popper's development of the notion of verisimilitude is in terms of the logical content of a theory being partly true and partly false. The class of non-tautological true statements in the logical content is the 'truth content,' and the class of (all) false statements, the 'falsity content.'[20] In the present section, however, an earlier notion of verisimilitude given by Popper will be considered, in which the truth content of a theory is composed of the class of all true statements in its logical content.[21] This should simplify present considerations without weakening Popper's position.

Popper's conception of verisimilitude may be clearly presented in terms of the Deductive Model. (In what follows it will be assumed that all statements of conditions or background knowledge C_1 to C_4 are true, and laws L_1 and L_2 are false.) L_1, in conjunction with statements of conditions C_1, entails true explananda E_1; and in conjunction with C_2, gives false explananda E_2. Similarly, L_2 in conjunction with C_3 subsumes true explananda E_3, and in conjunction with C_4 gives false explananda E_4. In this way E_1 and E_3 are the respective truth contents of L_1 and L_2, and E_2 and E_4 are their respective falsity contents. These relations may be schematised:

$$(23) \quad (L_1 \wedge C_1) \vdash E_1,$$
$$(L_1 \wedge C_2) \vdash E_2,$$
$$(L_2 \wedge C_3) \vdash E_3,$$
$$(L_2 \wedge C_4) \vdash E_4.$$

---

[19] Although, as discussed in Chapter 3, Popper believes that no statement in empirical science can be determined to be true, he feels justified in speaking in terms of a theory as having certain true consequences thanks to the work of Alfred Tarski, to whom he ascribes a correspondence theory of truth: cf. e.g. ibid., pp. 223ff., and Popper (1973), pp. 44ff. For a critique of Popper's interpretation of Tarski, see Haack (1976); for problems concerning the applicability of Tarski's conception in the case of the comparison of theories, see Kuhn (1970c), pp. 265f.

[20] Popper (1973), pp. 47ff.

[21] Cf. Popper (1962), pp. 391ff.

At first it might be thought that a measure of verisimilitude for L_1 might be determined by subtracting the number of statements in its falsity content E_2 from the number in its truth content E_1, and that a measure for L_2 might similarly consist in subtracting the number of statements in E_4 from that in E_3.[22] But since each of the sets of explananda should contain an infinite number of statements, this method cannot be used. Popper thus suggests rather that two theories L_1 and L_2 be defined as comparable with regard to verisimilitude only in case: (a) the truth content of one of them, L_1, is a proper subset of the truth content of the other, L_2, while the falsity content of L_2 is identical with, or a proper subset of, the falsity content of L_1; or (b) the truth content of L_1 is identical with, or a proper subset of, the truth content of L_2, while the falsity content of L_2 is a proper subset of the falsity content of L_1.[23] In either case we should say that L_2 has a greater verisimilitude than L_1.

Following (23), Popper's definitions may be schematised in terms of the true and false explananda of each of L_1 and L_2:

(24) $(E_1 \subset E_3)$ and $(E_4 \subseteq E_2)$ implies that $(VL_1 < VL_2)$,

(25) $(E_1 \subseteq E_3)$ and $(E_4 \subset E_2)$ implies that $(VL_1 < VL_2)$.

Aside from the fact that Popper gives no examples of scientific theories which he believes actually to be related in this way, Pavel Tichý has proved that on Popper's definitions no false theory can have a comparably greater verisimilitude than any other.[24] In the case of (24), Tichý has shown that if the truth content of L_1 were a proper subset of the truth content of L_2, then it could not be the case that the falsity content of L_2 be identical with, or a proper subset of, the falsity content of L_1. This would be so for the following reasons: Since the truth content of L_1 is a proper subset of the truth content of L_2, there is an element in L_2's truth content that is not in that of L_1. And since L_2 is false, there is at least one element in its falsity content. The conjunction of these two elements is false, and since each conjunct is in L_2, the conjunction itself is a part of L_2's falsity content.

---

[22] Cf. e.g. ibid., p. 234.

[23] Cf. Popper (1972), p. 52. These definitions were deleted from Popper (1973), but were not replaced with alternatives affording a clear conception of how one of two false theories can have a greater verisimilitude than the other.

[24] See Tichý (1974).

But it could not be a part of L_1's falsity content, for if it were, its true conjunct – just by being in L_1 and being true – would have to be in L_1's truth content, and this possibility has already been excluded. Thus it could not be the case that the falsity content of L_2 be identical with, or a proper subset of, the falsity content of L_1.

The above argument may be schematised:

(26) $(E_1 \subset E_3)$ and $(E_4 \subseteq E_2)$;
        thus, $(\exists x)(x \in E_3 \wedge \neg (x \in E_1))$;
        and, since L_2 is false,
        $(\exists y)(y \in E_4)$;
        thus, $(x \wedge y) \in E_4$;
        but $\neg ((x \wedge y) \in E_2)$, since $\neg (x \in E_1)$;
        thus, $\neg (E_4 \subseteq E_2)$.

Tichý uses a similar argument to do away with (25) above.

These arguments by Tichý, and similar ones by David Miller,[25] strike a serious blow to Popper's concept of verisimilitude. Though Popper might possibly get by these objections by having the notions of truth content and falsity content extend only to atomic statements and their negations, the efforts of Tichý and Miller have nevertheless succeeded in throwing a question mark on the concept of verisimilitude as a whole.

The considerations of the present chapter to this point have been concerned with the question of whether Popper has been able to avoid scepticism via the presentation of a conception of how one of two contradicting false theories may be thought to be a progression beyond the other, and to reveal how his attempts in this direction can be captured on the Deductive Model. It has been shown that his notion of content does not take account of theory conflict, and that his notion of verisimilitude is formally inadequate. We thus see that Popper has failed to afford a consistent conception of scientific progress that is in keeping with his view that successive theories contradict one another, and has consequently not succeeded in avoiding scepticism.

The fact that Popper has not provided a consistent conception of how one false theory might be a progression beyond another does not

---

[25] See Miller (1974).

mean that the provision of such a conception is impossible. But even if one were able to produce a clear account of theory succession which is in keeping with Popper's view, there exist problems of a more fundamental nature concerning the very applicability of the notions he employs in conceiving of science.

4. THE PROBLEMS OF MEANING VARIANCE AND THE NATURE OF SCIENTIFIC THEORIES

As presented in the previous chapter, the problem of meaning variance concerns the inability of accounts based on the Deductive Model to handle cases in actual science where terms such as "mass" undergo a change in meaning in their employment in different theories. Though Feyerabend has pointed to this problem in criticising the empiricist conception of theory succession, the fact that Popper's view is also based on the Deductive Model means that it arises here as well. Thus, just as saying that one theory entails another presupposes that their common terms have the same meaning, so does saying that one theory contradicts another. In other words, if there is a change in meaning of the terms common to two theories, then even if these theories were syntactically to contradict, this contradiction would be illusory. This in turn has relativistic consequences on the Popperian conception, for it implies that an observation statement purportedly corroborating one theory while falsifying another does not actually do so, since, given meaning variance, the terms it employs would not have the same meanings in both theories.[26] We see therefore that such a notion as Popper's concept of verisimilitude, even if it were successfully formulated in a consistent way, would fail to account for those cases in actual science where in the transition from one theory to another terms do change in meaning.[27]

The second problem to be mentioned here, that of the nature of scientific theories, is closely tied to the problem of meaning variance and to the empiricists' problems of theoretical terms and correspondence rules. At this point this problem will be expressed merely as a

---

[26] This problem has come to be called 'the paradox of meaning variance.' For attempts at its resolution from within the Popperian framework, see e.g. Giedymin (1970) and Martin (1971).

[27] For similar comments see Kuhn (1970c), pp. 234–235.

doubt, based on the existence of these other problems, as to whether actual scientific theories do have the form of universal or general statements having truth-values. In conjunction with this doubt, however, it may be pointed out that in actual science theories are often associated with *models* of one form or another, and neither Popper nor the empiricists have shown how such scientific models might be related to theories conceived as statements.[28]

In the next chapter Imre Lakatos' development of Popper's views will be considered, as will Popper's 'three requirements for the growth of knowledge.' In both cases it will be found that to the extent that Popper and Lakatos rely on the Deductive Model they do not avoid the criticisms of the present section, and that to the extent that they do not, they no longer explain science, but merely describe it.

---

[28] Though Ernest Nagel, for example, has provided an interesting discussion of the nature of scientific theories and the possible role played by models in them, he has not made a direct attempt to explicate how models might function in such theories when conceived as statements: see Nagel (1961), Ch. 6.

CHAPTER 6

POPPER, LAKATOS, AND THE TRANSCENDENCE OF THE DEDUCTIVE MODEL

1. SOPHISTICATED METHODOLOGICAL FALSIFICATIONISM

One criticism of Popper's view that is suggested in the writings of Thomas Kuhn is that in actual science a theory is never rejected unless there is another theory to take its place.[1] Imre Lakatos recognises a problem of this sort, but does not follow the path indicated by Kuhn, for he sees it as implying that theory change is an irrational process which can be analysed solely from within the realm of (social) psychology.[2] The alternative Lakatos thus chooses is to develop further Popper's conception in the context of the Deductive Model[3] in an attempt to "escape Kuhn's strictures and present scientific revolutions not as constituting religious [*sic*] conversions but rather as rational progress."

In the subsequent chapters of the present study however, it will be shown that the views of Kuhn (and Feyerabend) may be pursued in a natural way so as to eventuate in a conception in which the transition from one theory to its successor is based on both rational and empirical considerations. But in the present section Lakatos' conception – 'sophisticated methodological falsificationism' – will be analysed, and will be found to suffer from just that problem Lakatos ascribes to Kuhn, namely, the failure to provide rational grounds for preferring one of two competing theories to the other.

Lakatos' downfall in this regard is occasioned by his acceptance of Popper's arguments that in the testing of a theory one can never be

---

[1] Cf. Kuhn (1962), p. 77.

[2] Lakatos (1970), p. 93.

[3] That Lakatos' reasoning presupposes the Deductive Model is clear from the way in which he refers to it in his (1970), where he speaks of "the deductive model of the test situation" (pp. 116–117) and "our critical deductive model" (p. 131), and is further confirmed by his explicit use of it on pp. 185ff. of the same article. Quote following, ibid., p. 93.

sure of having established true statements of initial conditions.[4] In agreeing with Kuhn that a theory tested in isolation is not to be considered invalid on the basis of the results of its tests, Lakatos presents 'an imaginary case of planetary misbehaviour' in which Newton's gravitational theory resists successive attempts at falsification for the reason that the conditions of each test are found to be different from what they were thought to be at the time of making the test.[5] Thus, according to Lakatos, since we can never be sure of the truth-value of statements of initial conditions, we can never be sure that we have falsified a theory treated in isolation. But, as will be seen below, on the Deductive Model this reasoning applies not only to the present case, but also to that in which two theories are involved, and Lakatos does not succeed in showing falsification to be any more reasonable in the case of two theories than it is in the case of only one.

Lakatos presents his conception of the circumstances under which a theory may be considered falsified as follows:

For the sophisticated falsificationist a scientific theory T is *falsified* if and only if another theory T' has been proposed with the following characteristics: (1) T' has excess empirical content over T: that is, it predicts *novel* facts, that is, facts improbable in the light of, or even forbidden, by T; (2) T' explains the previous success of T, that is, all the unrefuted content of T is included (within the limits of observational error) in the content of T'; and (3) some of the excess content of T' is corroborated.[6]

This conception is essentially the same as Lakatos' notion of one theory having a higher degree of corroboration than another,[7] and, setting aside his parenthetical qualification, both notions can be schematised in terms of the Deductive Model most simply thus:

---

[4] This view was discussed in Chapter 3; and the problems it faces were set aside in Chapter 5 in order to make more reasonable Popper's attempts to provide a conception of progress. It will here be taken up again since it plays a central role in Lakatos' considerations. Cf. ibid., pp. 99ff.

[5] Ibid., pp. 100–101. It may be suggested here however that Lakatos is missing Kuhn's point. Kuhn is not denying that a theory treated in isolation may be seen to have flaws – rather, he is suggesting that a theory is not *rejected* unless there is an alternative to take its place. See e.g. Kuhn (1962), p. 77.

[6] Lakatos (1970), p. 116.

[7] Cf. Lakatos (1968), pp. 375–384.

$$(27) \quad (L_1 \wedge C) \vdash E_t \wedge E_f;$$
$$\qquad (L_2 \wedge C) \vdash E_t \wedge \neg E_f.$$

Here "L_1" is to denote theory T, which in conjunction with statements of conditions C implies certain true (or corroborated) empirical statements E_t, and certain false ones E_f. "L_2" denotes the superior theory T', which in conjunction with C entails the same true statements E_t, but instead of implying E_f implies $\neg E_f$.[8]

The above schematisation makes it clear that, on Lakatos' reasoning, just as a theory treated in isolation cannot be falsified, neither can it be falsified in the presence of another theory. And this is so for the same reason, namely, that we can never be sure that our statements of conditions C are true. Thus, as is evidenced by (27), if we cannot be sure of the truth-value of C, then we cannot know whether certain false statements are implied by one of the theories and not instead by the other. Though it might be argued that a more sophisticated schematisation than (27) is required in order to capture Lakatos' reasoning here, any such representation would also require the determination of the truth of C, and would thus lead to the same result. Thus we see that Lakatos has himself failed to provide a positive conception in which one theory is preferable to another on the basis of rational considerations.

Lakatos in fact recognises this problem, and devotes a good part of his paper 'Falsification and the Methodology of Scientific Research Programmes' to wrestling with it in various forms. For example, in treating of the problem in the context of crucial experiments (in which, on the Deductive Model, C must be determined true), he in the end suggests that, "we may – *with long hindsight* – call an experiment crucial if it turns out to have provided a spectacular corroborating instance for the victorious programme and a failure for the defeated one."[9] But it is easy to see that this is no rational criterion of

---

[8] At one point in his writings Lakatos actually makes a suggestion implying that we should here treat statements of conditions C as being a *part* both of T and T': see ibid., p. 382. While doing so would have no effect on the issues raised in the present discussion, it may nevertheless be pointed out that not only does this suggestion appear intuitively unsound, but Lakatos himself is not at all consistent in this regard. Cf. e.g. his above-mentioned 'imaginary case of planetary misbehaviour.'

[9] Lakatos (1970), p. 173. Next quote, Feyerabend (1970), p. 215; see also Kuhn (1970c), p. 262.

acceptability, for it does not tell us how long a hindsight we need; and even if it did draw the line at some point, that point would surely be arbitrary. As Feyerabend has said, "if you are permitted to wait, why not wait a little longer?" Lacking an answer to this question, Lakatos succumbs to the very charges of irrationalism that he levels at Kuhn. But this is not all, for the question Lakatos ought really to be answering here is not when we should call an experiment crucial, but when we should consider a research programme victorious. And if he is to answer this question at all, his response in the present context must be to the effect that the victorious programme is the one which, with hindsight, is supported by crucial experiments. But, as we see, this reasoning only leads him in a circle.

A consideration of the above discussion might lead however to the suggestion that to a large extent the critical points raised there do not directly affect Lakatos, but only show the Deductive Model to be unable to capture what he is saying. That the model does not fairly represent Lakatos is evident even in the case of his conception of falsification. There, for example, he speaks of part of the content of T being included, within the limits of observational error, in the content of T', and as has been admitted above, the Deductive Model does not capture this notion.

However, an attempt to conceive of Lakatos' position independently of the Deductive Model would not save it from the above criticism: on *any* interpretation (such as, e.g., Feyerabend's) he would have failed to provide a conception of how one theory is preferable to another on the basis of rational considerations. But, in any case, it would be naïve to deny that the Deductive Model is lying in the background of Lakatos' works discussed here, as is evident from such things as his conceiving of theories as sentences having truth-values, and his employing the Popperian concept of content, as well as his expressly mentioning the presence of the model in his considerations.[10] Thus in such cases as where he speaks of "the limits of observational error" we must see him as stepping outside the model in order to make his position appear more reasonable. This means, however, that these excursions into aspects of science which cannot be explained on the basis of the model are actually ad hoc, and while

---

[10] Cf. p. 41, n. 3, above.

they are consequently often correct, they must be relegated to the status of mere description, and be seen to lie outside what can be accounted for on Lakatos' philosophy of science.

2. POPPER'S 'THREE REQUIREMENTS FOR THE GROWTH OF KNOWLEDGE'

After discussing the notions of content and verisimilitude in Chapter 10 of *Conjectures and Refutations*, Popper goes on in the same chapter to suggest three requirements for the growth of knowledge. Popper's requirements, similar to a number of remarks made by Lakatos, seem for the most part to be quite reasonable, but they gain little support from his view of science based on the Deductive Model.

Popper's *first requirement* is that a new theory:

should proceed from some *simple, new* and *powerful, unifying idea* about some connection or relation (such as gravitational attraction) between hitherto unconnected things (such as planets and apples) or facts (such as inertial and gravitational mass) or new 'theoretical entities' (such as fields and particles).[11]

We might begin by investigating the extent to which Popper's philosophy of science supports our intuitive understanding of this requirement. That a new theory should proceed from some simple, powerful (logically strong) and unifying idea is, on Popper's view, all to say the same thing: that the theory have great content.[12] It may be pointed out, however, that Popper's notion of simplicity is itself thus counterintuitive, since it suggests that the simplicity of a theory should increase with each empirical statement conjoined to it. Furthermore, Popper's requirement must be a relative one – i.e. the new idea must have a great content relative to that of the idea in the earlier theory. But on Popper's view the old and new theories are to contradict, and as has been shown in the previous chapter, Popper has not provided us with a concept of content that can be applied in such cases. Thus this aspect of his first requirement is either vacuous, or

---

[11] Popper (1962), p. 241.
[12] Concerning 'simplicity' see e.g. Popper (1957b), p. 61; for 'logical strength' see e.g. Popper (1962), p. 219; and for 'unification' see ibid., p. 220.

else it quite transcends what Popper is warranted in demanding given his basis in the Deductive Model.

A second point that may be raised concerns Popper's suggestion that a new theory should proceed from a *new* idea. On the Deductive Model, however, the potential viability of a theory should depend on its formal structure, either considered in isolation or in relation to the structure of other theories. Thus, unless it is intended tautologically, this suggestion is quite antithetical to Popper's position in the logic of science. And, in any case, viewed simply on its own merits, this demand not only seems unreasonable, but is at variance with the history of science (cf. e.g. the revitalisation of the heliocentric conception, and the resurgence of atomism).

Thirdly, that Popper should speak of an *idea* as being distinct from the theory to which it leads is not at all indicated by his conception of theories as being universal statements. And lastly, Popper's reference to gravitational attraction, inertial and gravitational mass, and such theoretical entities as fields and particles only gives rise to the problems of theoretical terms and correspondence rules, which are confounded in Popper's case by his having no criterion for distinguishing between laws and theories.

Thus while Popper's first requirement is of some merit, it nevertheless cannot be said to be warranted on the basis of his own philosophy of science in which scientific theories are general statements of the sort suggested by the Deductive Model.

Popper's *second requirement* is that of independent testability:

[A]part from explaining all the *explicanda* which the old theory was designed to explain, [the new theory] must have new and testable consequences (preferably consequences of a *new kind*); it must lead to the prediction of phenomena which have not so far been observed.[13]

Popper points out that this requirement is necessary (on his view) in order that the new theory not be ad hoc, for, since he makes no essential distinction between theories and statements, the new theory could otherwise just as well consist of a series of statements describing presently known phenomena.[14] His requirement that the new theory must have *new* consequences can be captured on the Deductive

---

[13] Ibid., p. 241.
[14] Ibid., pp. 241–242.

Model if we simply take it as meaning that it must have *different* consequences – but this would not prevent it from being ad hoc. On the other hand, Popper's suggestion that the new theory must lead to the prediction of phenomena which have not previously been observed, while it does imply that the theory not be ad hoc, is again quite beyond what is warranted on the basis of his position in the logic of science. On the Deductive Model what is of importance are formal relations and the truth-values of such statements as those of initial conditions and explananda; *whether* certain of these truth-values are known is of course of relevance, but, from a formal point of view, *when* they are known is not. Thus we see here that not only is Popper unjustified in making this second requirement his own, but his conception of science suffers yet another setback through its inability to distinguish ad hoc theories from those that are not.

Popper's *third requirement* is that some of the predictions demanded by the second requirement should be verified, and that the new theory should not be refuted too soon.[15] Here we see that, in presupposing the second requirement, and in suggesting that the new theory not be refuted too soon, Popper's third requirement also rests on temporal notions that find no place in his philosophy of science. But in elaborating this requirement Popper says that "the successful new predictions which we require the new theory to produce are identical with the crucial tests which it must pass in order ... to be accepted as an advance upon its predecessor."

This further clarification may be seen as suggesting a criterion free of temporal considerations; but to the extent that it is so viewed, Popper's third requirement also fails to distinguish between innovative and ad hoc theories. Furthermore, Popper's (and Lakatos') view that statements of conditions can never be determined to be true – while counter-indicated by the employment of the Deductive Model – gives rise to the problem discussed in the previous section, namely, that we can then never be sure that the results of a test actually favour one theory and not perhaps the other. And, setting this problem aside – i.e. assuming that the truth-values of statements of conditions can be determined, Popper has still provided us with no consistent conception of how the superiority of one theory over another can be

---

[15] Ibid., p. 247; quote following, ibid.

based on crucial tests in the case where both theories – conceived as universal statements – are false.

Thus while Popper's third requirement, seen as suggesting that a superior theory pass certain crucial tests which its rival fails to pass, is quite reasonable and in keeping with a number of important episodes of the history of science, it at the same time lies beyond what Popper is justified in demanding from the basis of his own philosophy of science.

In reviewing the development of the present study to this point we see that both Popper's and the empiricists' attempts to understand the nature of science and scientific progress through the employment of the Deductive Model face a good many difficulties. Perhaps the most important of these are not internal problems, but problems of application. As will be argued in the next chapter, it is largely the existence of such problems as concern the application of the Deductive Model that has led Kuhn and Feyerabend to claim that in certain cases of actual scientific advance the theories involved are incommensurable.

CHAPTER 7

KUHN, FEYERABEND,
AND INCOMMENSURABILITY

1. 'INCOMMENSURABILITY' IN ITS NEGATIVE SENSE

The development of the logical empiricist and Popperian conceptions of scientific progress in terms of the Deductive Model has shown each to suffer a serious drawback: the empiricist view affords no conception of theory conflict, and the Popperian view provides no consistent conception of progress itself. Nevertheless, the intuitive notions motivating each of these philosophies of science, considered independently of the model, appear quite sound. One is still inclined to admit that, in some sense, succeeding theories do subsume their rivals, and that, in spite of this, such theories conflict with one another. Thus we might accept, for example, a description of theory succession in which the superior theory is said to explain both what its rival is able to explain, as well as certain of those states of affairs which are considered anomalous to the rival. But the problem here lies in the failure of the Deductive Model, and consequently the failure of both the empiricist and Popperian conceptions of science, to provide an account of this sort of phenomenon. And, as regards the employment of the model itself, we may add to this the problem of meaning variance, and the very question as to whether scientific theories have the form of universal statements, as the model suggests.

This distinction between the provision of an adequate description of science on the one hand, and the provision of an explanation of science (based on the Deductive Model) on the other, might help us to understand the attitude of Thomas Kuhn towards the Popperian philosophy of science. While Kuhn agrees with many of the claims made by Popper and Lakatos, especially where they suggest a non-cumulative conception of theory succession, he nevertheless feels that there is a basic difference between his and their ways of seeing science. For example, as regards Popper, Kuhn has expressed the conviction that "our intentions are often quite different when we say the same things;"[1] and, after likening their differences to a switch of gestalt, Kuhn asks, "How am I to show him what it would be like to wear my spectacles when he has already learned to look at everything I can point to through his own?" In another place Kuhn describes his single most fundamental difference from Popper as resting in the latter's belief that the problem of theory choice can be resolved in such a way as is suggested by the notion of verisimilitude – a belief which shares with the views of empiricists such as Carnap and Reichenbach the presupposition that "canons of rationality [are to] derive exclusively from those of logical and linguistic syntax."[2]

Now, it may be noted that Kuhn is not here, nor anywhere else, suggesting that science is not a rational enterprise. Quite the contrary: in keeping with a number of claims made by Popper and Lakatos he has in fact indicated more than once that he sees theory choice in science as being based on good reasons – reasons which are, furthermore, "of exactly the kind standard in philosophy of science: accuracy, scope, simplicity, fruitfulness, and the like."[3] But, from the point of view of the present study, where Kuhn does differ from Popper and Lakatos may be said to be in his view that the nature of theory succession in science, and with it the very nature of scientific rationality itself, does not have the form suggested by the employment of the Deductive Model.

This way of seeing the main difference between Kuhn's view and those of the empiricists and Popperians suggests that such charges of irrationalism as have been levelled at Kuhn by, for example, Lakatos presuppose the very basis Kuhn is calling into question. What seems to be the case is that many of Kuhn's critics presuppose, either

---

[1] Kuhn (1970b), p. 3; next quote, ibid. As regards Lakatos, see Kuhn (1971).

[2] Kuhn (1970c), p. 234. See also pp. 266–267 of the same article, and Kuhn (1974), p. 504: "My main and most persistent criticism of the recent tradition in philosophy of science has been its total restriction of attention to syntactic at the expense of semantic problems."

[3] Kuhn (1970c), p. 261. See also, e.g. his (1970a), p. 199. For an instance of the belief that Kuhn sees the decision between scientific theories as being irrational, see Laudan (1977), pp. 3f. and 141.

wittingly or otherwise, that an account of science as a rational enterprise must take the form suggested by either the Deductive Model or some other formal construction, and furthermore, that any alternative approach automatically lies outside what can correctly be called the philosophy of science.[4] But this view has nowhere been supported – let alone established – by argument; and, considering that the Deductive Model suffers so many failings, one wonders if it would not be wiser to accept Kuhn's view as an invitation to the development of an alternative, non-formal, conception of science as a rational (and empirical) enterprise.

The dissatisfaction expressed by Kuhn with regard to the empiricist and Popperian conceptions of science and scientific progress is largely shared by Paul Feyerabend. But where Kuhn approaches science mainly from the point of view of an historian, Feyerabend, while he does argue from historical examples, takes more the approach of a philosopher of science. Having early studied under Popper, Feyerabend was quick to see the failings of the empiricist view, pointing out its problem of meaning variance as well as emphasising its problem of consistency, as has been mentioned earlier in this study. And in his somewhat more recent work Feyerabend has also come to criticise the Popperian notion of verisimilitude for reasons pertaining to meaning variance, viz., that because of the differing concepts behind their common terms the respective contents of certain theories cannot be compared.[5]

Thus, when it comes to the question of the nature of theory succession in science, the present study leads to the view that the essence of both Kuhn's and Feyerabend's dissatisfaction with the empiricist and Popperian accounts may be seen as stemming from problems faced in the attempted application of the Deductive Model to actual science. And, to take this suggestion one step further, we may say that the respective claims of Kuhn and Feyerabend that in certain cases of theory succession the theories concerned are *incommensurable* amount to, *in their negative sense*, the claim that the Deductive Model (or, as might also be urged, any formal construction) fails to

---

[4] This view is still held by influential commentators today: see e.g. Stegmüller (1979), p. 69.
[5] Cf. e.g. Feyerabend (1970), pp. 220–222.

provide a realistic conception of the phenomenon as it actually occurs in science.

2. 'INCOMMENSURABILITY' IN ITS POSITIVE SENSE

The term "incommensurable" is defined in the *Concise Oxford Dictionary* as: "(Of magnitudes) having no common measure integral or fractional (with another); irrational, surd; not comparable in respect of magnitude; not worthy to be measured with." Kuhn and Feyerabend each borrow the term from this mathematical or mensural context and use it in attempts to express what is perhaps the same particular insight concerning the nature of the relation between certain scientific theories. It may be noted however that they do not apply the term to exactly the same cases. While both agree that such theories as Newtonian and Einsteinian mechanics are incommensurable,[6] where Kuhn would consider, for example, the geocentric and heliocentric systems of astronomy also to be so, Feyerabend would not.[7] Certain differences in their respective applications of the notion will be noted as this study proceeds, but at this point emphasis will be placed on what is common to their views, in an attempt to delineate one central positive aspect of 'incommensurability' capable of providing a starting point for the development of an alternative conception of theory succession in science.

On almost every occasion in which Kuhn speaks of incommensurability, and at a number of places where Feyerabend does so, those points of view which are taken to be incommensurable are at the same time also seen as being competing or incompatible.[8] Keeping this in mind we may also note that, as is suggested by Feyerabend's critique of empiricism, an important rationale for the claim that certain theories are incommensurable is the suggestion that there is a change in the meaning of some or all of the 'descriptive' terms common to the theories in question when one moves from one of

---

[6] Cf. e.g. Kuhn (1962), p. 102, and Feyerabend (1965b), pp. 230ff.

[7] For Kuhn, cf. his (1962), p. 102; Feyerabend does not include this case among his examples in, e.g., his (1975), pp. 276–277, and has pointed out in a personal communication that he does not take it to constitute an instance of incommensurability.

[8] Cf. e.g. Kuhn (1962), pp. 103 and 165, and (1970a), p. 175; and see Feyerabend (1975), p. 274.

them to another. For example, in considering the transition from classical celestial mechanics to general relativity Feyerabend has said, "the meanings of all descriptive terms of the two theories, primitive as well as defined terms, will be different,"[9] and, speaking more generally, Kuhn has said, "In the transition from one theory to the next words change their meanings or conditions of applicability in subtle ways."[10]

As has been noted in earlier chapters, the viability of such claims concerning meaning change creates important problems as regards the applicability of the Deductive Model, and leads directly to relativistic consequences in the case where theories are seen to contradict. Unfortunately, however, it seems that those who have dealt with this problem have assumed that the Deductive Model or some similar basis employing the first-order predicate calculus is the only possible vehicle for the analysis of theory change in science. As a result of this, the relativism that arises in such contexts has been taken to imply a relativism in Kuhn's and Feyerabend's claims themselves; and it has not been realised that in this particular case, granting that such changes in meaning do occur, it is rather the Popperian conception that leads to relativism, as has been suggested in Chapter 5.

Though both Kuhn and Feyerabend have in their earlier writings based their claims that certain theories are incommensurable largely on this change of meaning of individual terms, neither of them has identified incommensurability with meaning variance. In fact, in reaction to his critics' coming more or less to equate these two notions, Feyerabend has said such things as that "in the decision between competing theories, 'meanings' play a negligible part,"[11] and that "As far as I am concerned even the most detailed conversations about meanings belong in the gossip columns and have no place in the theory of knowledge." Both Kuhn and Feyerabend have emphasised rather that the shift between incommensurable theories involves a *fundamental* change, and in a recent work Feyerabend has instead begun his discussion of incommensurability by taking it to be

---

[9] Feyerabend (1965b), p. 231.

[10] Kuhn (1970c), p. 266. Note that Kuhn does not claim that all descriptive terms should change in meaning. In this regard cf. also p. 506 of (the discussion following) Kuhn (1974).

[11] Feyerabend (1965a), p. 267; next quote, Feyerabend (1965b), p. 230.

exemplified by gestalt-switch phenomena.[12] In this same vein, Kuhn in his earlier writings has suggested that "the switch of gestalt, particularly because it is today so familiar, is a useful elementary prototype for what occurs in full-scale paradigm shift,"[13] and furthermore that "Just because it is a transition between incommensurables, the transition between competing paradigms cannot be made a step at a time, forced by logic and neutral experience. Like the gestalt switch, it must occur all at once (though not necessarily in an instant) or not at all."

Following this lead, the essence of the notion of *incommensurability – in its positive sense –* will here be taken to be the idea of being related as are the different aspects of a gestalt-switch figure. In what follows of this study this idea will be developed so as to provide a general characterisation of theory change which is applicable to any case of competing scientific theories, and which not only avoids the problem of relativism that meaning change gives rise to on the Deductive Model, but is capable of explaining both theory conflict and scientific progress.

---

[12] Feyerabend (1975), pp. 225ff.
[13] Kuhn (1962), p. 85; next quote, ibid., p. 150.

CHAPTER 8

THE GESTALT MODEL

1. A MODEL OF THEORY CHANGE VS. AN EXAMPLE OF PERCEPTUAL CHANGE

Gestalt-switch diagrams have perhaps most often been used as paradigmatic examples of entities which can be perceived in completely different ways without their changing, and without there being a change in the perceiver's physical relation to them. N. R. Hanson, pursuing Wittgenstein's remarks concerning seeing and 'seeing as,'[1] has employed gestalt-switch figures in this sort of way in considering cases relevant to the philosophy of science. In the chapter of his book *Patterns of Discovery* entitled 'Observations,' for example, he has followed this line of thought, suggesting that where "Tycho and Simplicius see a mobile sun, Kepler and Galileo see a static sun."[2]

Kuhn, though he is aware of the difficulties involved in the idea that those employing different theories need *see* the world differently,[3] nevertheless devotes Chapter X of *The Structure of Scientific Revolutions* to a development of his own thoughts along this line. Feyerabend too is attracted to the idea, but feels he cannot accept it: "I strongly sympathize with [Hanson's] view, but I must now regretfully admit that it is incorrect. Experiments have shown that not every belief leaves its trace in the perceptual world and that some fundamental ideas may be held without any effect upon perception."[4]

But if we follow Hanson further, we see that in the chapter of his book entitled 'Theories' he employs the gestalt-switch phenomenon not so much as an example at the level of observation, but as an

---

[1] Cf. Wittgenstein (1953), IIxi.

[2] Hanson (1958), p. 17. It may be noted that this is perhaps the most extreme thing that Hanson says in this regard, and that he devotes more space to a discussion of the differences between a layman's and a scientist's ways of seeing a piece of scientific apparatus.

[3] Cf. Kuhn (1962), pp. 85 and 113ff.

[4] Feyerabend (1965b), p. 247. See also Feyerabend (1977), p. 365n.

analogue to the level of theory. Here, in the context of criticising accounts based on the Deductive Model, he argues by analogy from a gestalt-switch figure that there is an important distinction between observational detail statements and the conceptual frameworks (theories) within which they are cast.[5]

Jan Andersson and Mats Furberg, employing the same gestalt-switch figure as Hanson, have obtained similar sorts of results which, though intended to provide insight regarding the nature of perception, may be seen from the point of view of the present study to provide the outline of a *model* of how different theories relate one to another.[6]

2. THE DUCK-RABBIT AS A MODEL OF THEORY SUCCESSION IN SCIENCE

It is intended that the term "model" as used here have the meaning it usually has when employed in everyday and scientific contexts – a meaning which is closely related to the notion of analogy. Whether what follows be called a model or an analogy is thus not of great concern to the present study, but seeing as the gestalt-switch figure to be treated below is being advanced especially to serve as the intuitive basis of the account of the next chapter, and is not being introduced after the fact in order to support certain singular conclusions, it seems more in keeping with ordinary usage to call it a model.

---

[5] Cf. Hanson (1958), p. 87: "Consider the bird-antelope in fig. 12. Now it has additional lines. Were this flashed on to a screen I might say 'It has four feathers.' I may be wrong: that the number of wiggly lines on the figure is other than four is a conceptual possibility. 'It has four feathers' is thus falsifiable, empirical. It is an observation statement. To determine its truth we need only put the figure on the screen again and count the lines.

Fig. 12

"The statement that the figure is of a bird, however, is not falsifiable in the same sense. Its negation does not represent the same conceptual possibility, for it concerns not an observational detail but the very pattern which makes those details intelligible."

[6] Andersson's and Furberg's treatment of the gestalt-switch phenomenon thus differs from the one to follow in the present chapter in that they do not consciously present it as a basis for an alternative conception of science: cf. Andersson & Furberg (1966), Chh. 5–7.

The following presentation of the Gestalt Model will thus consist in the pointing out of certain features of a gestalt-switch figure – features which in the next chapter will be seen to provide a foundation for the explanation of many important aspects of the relation between actual scientific theories. In contradistinction to the use of such a figure in the context of discussing certain perceptual phenomena, emphasis will here be placed on particular *conceptual* features that it exhibits.

What follows of this chapter thus constitutes only a presentation of the model, and not its employment. But it may be kept in mind that the remarks to be made here are intended ultimately to have relevance to full-fledged theories – theories such as Galileo's terrestrial physics and Newton's theory of gravitation.

The Gestalt Model differs from the Deductive Model in regard to flexibility. Any gestalt switch phenomenon may be used as a basis for considering the nature of the relation between particular theories. One may use, for example, Hanson's bird-antelope, or Wittgenstein's duck-rabbit[7] – whatever gestalt-switch figure best suits the peculiarities of the theories to be considered. In this study the duck-rabbit will be taken as the model, since it is capable of affording a starting point for a relatively general account of the nature of the relation between successive theories.

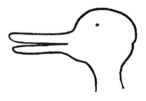

Figure

We might begin by asking what it is that is peculiar about the above figure so as to justify our calling it a gestalt-switch figure, and thus take it as differing in a special way from more ordinary drawings, such as Wittgenstein's 'picture-face.'[8] While it may be difficult

---

[7] Wittgenstein (1953), p. 194.
[8] Ibid.

to describe the difference in a way which everyone would find acceptable, we might try by suggesting that where the picture-face can be seen only in one way, the duck-rabbit can be seen in two.[9] And to this we might add that, as a matter of psychological fact, the duck-rabbit cannot be seen both as a picture-duck and a picture-rabbit by one person at one and the same time, in the sense in which, say, a picture of a red ball can be seen at one time to be of an object which is both red and round.[10]

3. THE SEEING OF AN ASPECT AS THE APPLICATION OF A CONCEPT

The shift from seeing the duck-rabbit as a picture-duck to seeing it as a picture-rabbit, or vice versa, Wittgenstein calls a 'change of aspect.' Each aspect is quite general in the sense that it exhausts the whole of the figure, and yet each is also exclusive of the other. In what follows the seeing of the duck-rabbit as a (picture-)duck[11] (or seeing it under the duck aspect) will be taken to be the application of the concept 'duck' to it, and thus here we may say that the shift from applying the concept 'duck' to applying the concept 'rabbit' is a fundamental and revolutionary one – the figure is seen completely differently once the other concept is applied.

If, upon first looking at the duck-rabbit, a person were able only to see the duck aspect, he could be helped to see the rabbit aspect by being told that what for the duck is its bill, for the rabbit is its ears, and so on. We would not argue or try to prove that the figure can be

---

[9] This has sometimes been misleadingly expressed by saying that gestalt figures are ambiguous. But where we might say of some linguistic entity such as a phrase or sentence that it is ambiguous, it would at least be metaphorical to say the same of gestalt figures.

[10] As expressed by Kuhn: "Though most people can readily see the duck and the rabbit alternately, no amount of ocular exercise and strain will educe a duck-rabbit." (1977), p. 6.

Note also that a treatment of the question as to *why* the duck-rabbit has the particular features it does is perhaps best handled by a psychologist; at any rate, such a treatment lies beyond the scope of the present study.

[11] Following Wittgenstein, in this study such terms as "picture-duck" will often be dropped in favour of the simpler formulation "duck." In all cases however these terms are meant to have reference only to the duck-rabbit figure.

seen as a rabbit; instead we would try to *show* the person the rabbit aspect. A second way of doing so, rather than 'translate' from the duck aspect as above, would be to translate from a more neutral description of the figure itself, telling the person for example that the long and narrow part of the figure is for the rabbit its ears, and that the indentation on the right side is its mouth, etc. Or a third way might be to use ostension, and simply point out parts of the rabbit aspect, naming them as we do so.

Once the person has seen both aspects he might easily compare them; for example, he could say that the duck is looking to the left, while the rabbit is looking up, or to the right.[12]

4. SIMULTANEOUS APPLICATION

As mentioned above, as a matter of fact the duck-rabbit cannot be seen both as a duck and as a rabbit at the same time by the same person; or, in other words, one person cannot *simultaneously* apply the concept 'duck' and the concept 'rabbit' to the duck-rabbit.[13] However, the two concepts can of course be applied simultaneously if each is being applied by a different person; and, as will be treated below, this is also the case if each concept is being applied to a different figure.

5. ASPECTUAL INCOMPATIBILITY

Perhaps the feature of the Gestalt Model which is the most difficult to describe is also that which makes gestalt-switch figures themselves so interesting, namely, the *conflict* that ensues when one attempts to see both aspects of such a figure at the same time. What is the nature of this sort of conflict or incompatibility, and can it be described in terms with which we are familiar independently of the existence of gestalt figures themselves?

---

[12] Cf. Andersson & Furberg (1966), p. 49: "the antelope head looks to the right, the pelican head to the left; the antelope head but not the pelican head has clearly distinguishable horns; etc."

[13] Similarly, Wittgenstein says, though with different emphasis: "But the *impression* is not simultaneously of a picture-duck and a picture-rabbit." (1953), p. 199.

If we compare the Gestalt Model and the Deductive Model on this point, we see that the sort of incompatibility that the Deductive Model can be used to express is essentially that of one sentence or statement negating another; in other words, conflict is there depicted as being of a linguistic nature, and as basically consisting in one law contradicting or denying what is entailed by another (plus statements of conditions). But in the present case we are not dealing with the making of statements or assertions so much as with the application of concepts. Upon seeing the duck-rabbit as a rabbit, one does not say, even to oneself, 'this is a rabbit'; rather, one simply applies the concept 'rabbit' to the figure. The sort of conflict that can arise here is prior to the employment of language; i.e., we might say that the conflict, while involving concepts, is pre-linguistic, for one's inability to see the duck-rabbit as both a duck and a rabbit at the same time is quite independent of his having words such as "duck" and "rabbit" to employ in describing the two aspects.

However, by continuing to speak in terms of the application of concepts we can come somewhat closer to a neutral description of the sort of conflict involved here. To this end then we should say that the concepts 'duck' and 'rabbit' are incompatible when the attempt is made to apply them simultaneously to the same thing. But even working at the more abstract level of concepts, we find that the situation cannot correctly be described as being essentially the same as that involving contradiction. This is so because a *contradiction* consists in the simultaneous affirmation and denial of one and the same proposition,[14] and not only are there no propositions or statements involved in the present case, but neither is there a notion of denial or negation. Rather, what we have here is more similar to what Aristotle would consider to be *contrary qualities* where, for example, he says, "nothing admits contrary qualities at one and the same moment."[15] But, in order to avoid misunderstanding, the sort of conflict considered here will simply be termed 'aspectual incompatibility,' and

---

[14] This notion of contradiction should be that which philosophers generally intend when they employ the term. Cf. e.g. Aristotle *De Int.* 17ª 31–33: "every affirmation has an opposite denial, and similarly every denial an opposite affirmation. We will call such a pair of propositions a pair of contradictories."

[15] *Cat.* 5ᵇ 39–6ª 1. Cf. also *De Int.* 24ᵇ 8–9, *Metaphy.* 1018ª 25–26, and *Post. An.* 88ª 28–29.

will be understood as involving the simultaneous application of differing concepts to one and the same thing, but not as involving the use of propositions, nor, consequently, as being the result of one proposition or statement negating or denying what is asserted by another.

6. THE UNIQUENESS OF THE REFERENT

In order for conflict to arise in the case of the duck-rabbit not only is it necessary that the concepts 'duck' and 'rabbit' be applied at the same time, but, as mentioned earlier, both must be applied to the same figure or thing. Thus, for example, if we had two duck-rabbit figures side by side we might simultaneously see one of them as a duck and the other as a rabbit. In such a case there would be no conflict. Expressing this in terms of concepts we might thus say that conflict arises in the present case only if it is the *intention* of the person employing the concepts that both be applied to the same thing, and that when this is so both may be said to have the same *referent*. In other words, we can say that the concepts involved are not incompatible in and of themselves, but that they conflict only when they are given the same *reference*.[16]

7. PREDICATES OF THE SAME CATEGORY

The fact that two differing concepts or predicates be simultaneously given the same reference is however not itself sufficient for conflict to arise, since we might see one and the same thing as being, for example, both red and round; i.e., we can apply the predicates 'red' and 'round' to the same thing without conflict ensuing. On the other hand, we cannot see (the whole of) one thing as simultaneously being both red and blue. The difference between these two cases will here be expressed by saying that in the latter case the predicates 'red'

---

[16] Note that this usage of the terms "referent" and "reference" is independent of any linguistic connotation they might normally have.

and 'blue' are of the *same category* 'colour,'[17] whereas in the former case the predicates concerned were respectively of the categories 'colour' and 'shape' or 'form.'

Employing this line of thought in the case of the duck-rabbit we should thus say that the concepts or predicates 'picture-duck' and 'picture-rabbit' are of the same category 'shape,' and that their being of the same category is a prerequisite for their conflicting when simultaneously applied to the duck-rabbit.

8. RELATIVE ACCEPTABILITY: ACCURACY, SCOPE AND SIMPLICITY

Once a person has been able to see the duck-rabbit both as a duck and as a rabbit, the question might arise as to which concept is the better applicable.[18] However, assuming that this person already has a definite view on the matter and considers e.g. the rabbit concept for some reason to be superior in this regard, if he were to attempt to convince another person who has only been able to see the duck aspect that this is so, the ensuing debate might well be characterised as a 'talking at cross purposes.' For example, what the first person calls the rabbit's ears, the second sees only as the duck's bill, and he might thus consider all the other's talk about 'ears' as being utter nonsense.

Accuracy

On the other hand, if the two parties are each able to see both aspects of the duck-rabbit, a meaningful argument might follow regarding whether the figure is more like a duck than a rabbit or vice versa. On first consideration it might appear that there is nothing in the figure to favour either aspect; but a closer inspection reveals that

---

[17] In this study categories will be treated as abstract entities, i.e. as entities which are of the sort to which predicates or concepts, rather than properties, belong; thus we will speak of the category: 'colour,' rather than of the category: colour.

[18] As expressed by Andersson and Furberg: "Which way of seeing fig. 3 is the more correct, that of the person seeing it as an antelope, or that of the one seeing it as a pelican?" (1966), p. 54.

the indentation which is the rabbit's mouth does not at all seem in keeping with the concept of the duck-rabbit as a duck.[19]

Assuming that we are unable to find anything in the figure which can be accounted for in the duck aspect while being anomalous to the rabbit aspect, we might then be inclined to say that the rabbit concept is the more *accurate*. Any part of the figure which may be criterial for the application of the concept 'duck' to the duck-rabbit is also criterial for the application of the concept 'rabbit'; but there is 'evidence' that favours the application of the rabbit concept that is also evidence against the application of the duck concept. Thus, though the two concepts can conflict prior to the discovery of something anomalous to one but not to the other, it is only with such a discovery that they can be judged as to their relative accuracy.

The indentation which is crucial to the determination of the relative accuracy of the two concepts might not have been visible to the naked eye. In such a case, the person advocating that the figure is more rabbit-like than duck-like might have suggested trying to find the notch (the rabbit's mouth) by using an instrument such as a magnifying glass. The unavailability of such an instrument would, in this case, have precluded the determination of either concept being more accurate than the other.

Scope

We might also imagine for the moment a slightly different case in which the figure contains not only the duck-rabbit head, but also the outline of a whole body. And let us say that one of the two concepts is clearly better applicable to the whole of this larger figure. (Note that the duck-rabbit is not to be thought of as actually *being* a figure of a duck as versus that of a rabbit, or vice versa. For present purposes the figure might just as well have been created by chance; and

---

[19] Cf. also ibid., p. 55: "The correctness of seeing a phenomenon under a certain aspect is supported then, if new perceptions of other parts of the phenomenon (or of the same parts but in new ways) strengthens the phenomenon's falling under the concept in question.

"When the seeing of an aspect allows a perceptual filling out in this way, its correctness can be tested in about the same way as can that of a scientific theory." Note however that here reference is being made to scientific theory in order to help clarify the nature of seeing under an aspect, and not vice versa, as in the present study.

the questions being raised here simply concern the relative applicability of two particular concepts to it.) In this case then we should
say that the concept which is the better applicable to the larger figure
has a greater *scope* than does the other, or, we might say that it is the
more *general* concept of the two.

Simplicity

But even granting the existence of the indentation, the supporter of
the duck concept might have rejoined by saying that the duck-rabbit
nevertheless closely resembles a duck – one which has suffered a
blow to the back of its head! But if this were all he had to say, his
claim would have to be recognised as being merely ad hoc, for,
where the notch is integral to the rabbit concept (as being the rabbit's
mouth), in the duck concept it is, in a sense, something 'added on.'
Thus while the concepts 'rabbit' and 'duck that has been struck in
the back of the head' might be equally accurate in accounting for
why the figure appears as it does, other things being equal the former
concept should be said to be the *simpler* of the two.

Should this mean then the abandonment of the duck concept? Not
at all. Though the rabbit concept may be the more accurate, it might
be pointed out that, for example, on first glance the figure still *looks*
more like a duck.

Similarly, the discovery of the anomalous notch *prior* to the seeing of the duck-rabbit as a rabbit should not have led to the rejection
of the duck concept (though it might well have led to one's wonder-
ing whether there exists another concept in which the notch plays an
integral role). To conceive of the figure as a duck with an unex-
plained indentation in the back of its head is better than to have no
clear concept of it at all.

Further scrutiny of the duck-rabbit reveals a bump behind the
rabbit's ears (under the duck's bill) which we might here consider
anomalous to both aspects. Again though, the presence of an aberra-
tion should not lead to the elimination of either concept.

The duck-rabbit is a figure which may be seen either as a picture-
duck or as a picture-rabbit, and is not simply a picture of a duck and
not that of a rabbit, or vice versa. Neither of the aspects in which the
figure may be conceived can be determined in any absolute sense to

be better than the other, and neither should be discarded. Though there is a sense in which the rabbit aspect both is superior to and subsumes the duck aspect, this superiority does not rest in the latter's being formally deducible from, or reducible to, the former; and though there is a sense in which the duck aspect is incompatible with the rabbit aspect, this incompatibility is not essentially one of contradiction.

At the same time as affording definite non-linguistic conceptions of conflict (as aspectual incompatibility) and progress (as relative accuracy, scope and simplicity), the Gestalt Model provides a positive conception of incommensurability. The Gestalt Model thus differs from the Deductive Model in the above sorts of ways, and as will be seen below, it differs further in that it can function as the basis of a conception of theory relations which is capable of handling meaning variance between theories, and which provides a more realistic conception of the nature of scientific theories themselves.

CHAPTER 9

THE PERSPECTIVIST CONCEPTION OF SCIENCE

1. THE PERSPECTIVIST CONCEPTION AS BASED ON THE GESTALT MODEL

In being based on the Deductive Model, the Popperian and empiricist views of science take scientific laws and theories to be statements of the form: for all x, if x has the property F, then x has the property G. The first step is conceptually to delineate a universe of x's having property F, and then it is to be empirically determined whether such x's also have property G. A number of these x's are thus to be observed and found either to have or not to have this property. There is no middle way – on this view either the predicate G is applicable or it is not, and if not, the law or theory is considered false.

In the present account quite a different tack is taken. First, a distinction is made between scientific laws and theories; and, with regard to *theories*, rather than start conceptually with a universe of objects which are to have some particular property, one may better be thought of as starting with some property, and then going on to find out in what sorts of instances it is applicable. In certain cases it may be discovered to fit quite well; in others, not so well. But the sharp distinction made in the logic of science between a predicate's either being applicable or not being applicable is here dropped.[1] On the present view a predicate may be more or less applicable, either by itself or as relative to some other predicate.

So here scientific theories are not taken to be statements having a truth-value, but are instead likened to individual empirical concepts or predicates which are intended to apply to certain phenomena – predicates such as 'red' or 'round,' or, with reference to the Gestalt Model, concepts such as 'picture-duck' and 'picture-rabbit.' Thus theories are here conceived of as being intended to apply to certain

---

[1] This difference thus serves also to distinguish the present view from the set-theoretic conception – cf. Chapter 11 below. It is also relevant to the notion of *box-thinking* taken up in Appendix VIII (p. 265&n., below).

states of affairs, and to be such that they may be judged to be more or less successful in their application.[2]

The motivation for this alternative approach is the desire to provide a conception of science differing from the empiricist and Popperian views in that it is able to account both for theory conflict and for scientific progress, and is also able not only to handle the problems of meaning variance and the nature of scientific theories, but to explain the way in which actual theories might well be incommensurable. Also, though the present view has gained much inspiration from the respective works of Kuhn and Feyerabend, it is not being presented as a direct reconstruction of the views of either of them; and, in fact, the notion of incommensurability to be treated here differs essentially from that described by Feyerabend as recently as in his *Science in a Free Society* (1978). Furthermore, the present conception is not directly concerned with the psychological or sociological aspects of science, but is to fall wholly within the realm of what is normally considered to be the philosophy of science. It might also be mentioned here that the present view is not at all intended to be prescriptive, or to provide 'methodological rules' for science or any other epistemological activity. Rather, it is intended to afford a realistic conception of central philosophically interesting aspects of science – a conception in which the move from one scientific theory to its successor is seen as being based on both rational and empirical considerations. Nor is the conclusion drawn from this that science itself provides the most reasonable means of gaining knowledge about reality, or that it is for any other reason an enterprise which ought to be pursued. In this study such questions are left completely open.

The basis for the alternative conception to be presented below is, in this study, being taken as the Gestalt Model. It may be noted however that a different starting point might have been taken, such as, for example, a comparison of colour concepts, or a direct treatment of

----

[2] Cf. Kuhn (1962), p. 147: "All historically significant theories have agreed with the facts, but only more or less. There is no more precise answer to the question whether or how well an individual theory fits the facts. But questions much like that can be asked when theories are taken collectively or even in pairs. It makes a great deal of sense to ask which of two actual and competing theories fits the facts *better*."

scientific theories themselves.[3] It is because so many features of gestalt-switch phenomena may be seen to have counterparts in the comparison of rival scientific theories that the heuristic value of these phenomena has resulted in their here attaining the status of constituting the basis for the present view. The following presentation will thus parallel the presentation of the Gestalt Model in the previous chapter, but certain points to be made will occasion the use of other sorts of examples and analogies.

2. SCIENTIFIC THEORIES AS CONCEPTUAL PERSPECTIVES

In the Gestalt Model the seeing of an aspect of a gestalt figure has been characterised as the application of a concept to the figure. In the present case then, a scientific theory will be understood as being in important respects analogous to a concept intended to have such an application. This notion of theory is to be as close to ordinary usage as possible, and thus should be largely in keeping with its definition in the *Concise Oxford Dictionary* as a "supposition or system of ideas explaining something, esp. one based on general principles independent of the facts, phenomena, etc. to be explained." This definition suffers however for its ignoring the hypothetical nature of theories – to be a theory a conceptual construct need not *succeed* in explaining what it is intended to explain. In what follows the term will thus be used in its more generally accepted sense, such that a scientific theory is an entity *intended* to have application to certain empirical states of affairs – an intention which may or may not be fulfilled.

Of course scientific theories seem intuitively to be much more complex than individual concepts, and the notion of a single empirical concept is here meant only to be employed as an analogue to that of a scientific theory. But scientific theories and empirical concepts

---

[3] The present conception as a matter of fact originated independently of the Gestalt Model, and was introduced in an earlier study via an example in which the different systems of co-ordinates formulable within the special theory of relativity were each taken to constitute a 'conceptual perspective.' Thus not only might the particular account of science being presented here have taken a different starting point, but the Gestalt Model itself is by no means intended to function as the basis of a philosophy of a more general nature.

intended to apply to certain states of affairs have sufficient in common that they may be grouped and discussed under the one heading: *conceptual perspective*. Thus where accounts based on the Deductive Model take scientific theories to be statements of the form: $\forall x \; (Fx \rightarrow Gx)$, here they are taken to be conceptual perspectives; and it is to the task of clarifying this notion, and of revealing in a general way its relevance to actual science, that the remainder of this chapter is devoted.

So an applied empirical concept may thus be thought to be a simpler form of conceptual perspective, and a scientific theory a more complex one. In the case of scientific theories a conceptual perspective may consist of a number of abstract notions, related in clearly defined ways. (In the next chapter the nature of the internal structure of scientific theories will be treated in greater detail.) More generally, a conceptual perspective of the more complex form is a conceptual system or framework which provides a structure in which one's thoughts about some particular aspect of the world can be organised. And while the interest of the present study is in scientific theories, there is no reason why we could not in a different setting consider, for example, such geometrical theories as those of Euclid, Lobachewsky, and Riemann each to constitute a conceptual perspective being applied to geometrical space.

With regard to the Gestalt Model we see then that the application of a scientific theory is to bear essential similarities to the phenomenon of seeing under an aspect,[4] and that just as the move from one gestalt concept to the other involves a fundamental change, so does the move from one scientific theory to its successor. Both may be said to involve a 'shift of conceptual perspective' – a shift which does not preclude a meaningful comparison of the perspectives concerned.[5]

---

[4] Cf. Hanson (1958), p. 90: "Physical theories provide patterns within which data appear intelligible. They constitute a 'conceptual Gestalt'."

[5] Cf. Kuhn (1976), pp. 190–191: "Most readers of my text have supposed that when I spoke of theories as incommensurable, I meant that they could not be compared. But 'incommensurability' is a term borrowed from mathematics, and it there has no such implication." For a recent instance of a misunderstanding of Kuhn in this regard, cf. Laudan (1977), p. 143.

It may be thought though that where in the Gestalt Model one can easily shift back and forth between perspectives, this is not so in the case of scientific theories. But, while it may take some time before a scientist grasps or *understands* a new theory, the actual shift from not understanding to understanding, a step which may occur more than once before the new theory or perspective is internalised, is clearly similar to the dawning of an aspect which occurs in the Gestalt Model. It may be noted also that such an understanding is not gained as the direct result of argument or critical discussion,[6] but on the contrary requires an uncritical posture on the part of the individual seeking to understand. And, once the scientist has understood both theories, though he might believe one of them to be clearly superior to the other, he can nevertheless easily shift from working within one to working within the other, just as a person can easily shift from one aspect of the Gestalt Model to the other.

3. LOGICAL SIMULTANEITY

The Gestalt Model also provides a notion of simultaneity that is of relevance in the case of successive scientific theories. Thus, just as the concepts 'duck' and 'rabbit' cannot be applied simultaneously to the duck-rabbit; we should say that competing scientific theories cannot be applied simultaneously.

Other authors have employed a notion of simultaneity similar to that intended here. For example, in the context of discussing the nature of incommensurability, Feyerabend has done so in saying that: "we may suspect that the family of concepts centring upon 'material object' and the family of concepts centring upon 'pseudo-after-image' are incommensurable in precisely the sense that is at issue here; these families cannot be used simultaneously ...."[7]

---

[6] For a similar point of view, see Feyerabend (1975), p. 229.

[7] Ibid., pp. 228–229. For another use of the term "simultaneity" in the context of the incommensurability question, see Stegmüller (1979), p. 69.

The concept of simultaneity given above is also potentially applicable to the case of complementarity in quantum physics, as is suggested e.g. by Pascual Jordan's saying: "The properties connected with the wave nature of light on the one hand and those connected with its corpuscular nature on the other ... can never appear in one and the same experiment at the same time" (1944), p. 132.

Part of what is of interest in considering the present notion of simultaneity is that it reveals a particular sense in which two distinct theories are to be independent of one another which is essentially different from the sense in which they might be conceived to be independent on a formal approach. From a formal point of view, a consideration of the relation between two theories, even if it is taken to be that of contradiction, requires that both theories be conceived as constituting the whole or part of one and the same formal system. But on the present conception, distinct theories are seen to be such that they cannot be applied simultaneously, and thus in principle to be such that, in application, they cannot be part of the same system. And this in turn means that when they are being treated as applied systems, formal relations cannot be established between them.

This being the case however does not preclude the determination of formal relations when the theories concerned are considered independently of their applications, i.e. when they are considered as purely formal systems. But it does suggest that the realisation of such relations might well be misleading, if for no other reason than that it perforce ignores the fact that the systems are independent in the above sense. Thus, in keeping with the negative sense of 'incommensurability' given in Chapter 7, we should here say that, as *applied* systems, competing scientific theories cannot be treated *logically simultaneously*.

4. PERSPECTIVAL INCOMPATIBILITY

As is suggested by the Gestalt Model, on the present account theory conflict is not conceived to be essentially of a linguistic nature, nor to involve the express denial or negation of (part of) one theory by another. More generally, the sort of conflict of interest here – *perspectival incompatibility* – is independent of whatever formal or syntactical relations might hold between the theories; i.e. it is independent of how the theories are related to each other when considered simply as unapplied systems. Rather, theory conflict is here conceived to arise in the *application* of the theories, each of which is intended to provide a positive account of one and the same realm of phenomena, and this conflict is to involve each of the theories as a whole.

The Relevance of the Incompatibility of Colour Concepts

As outlined above, both single empirical concepts and more complex empirical systems, when applied, constitute conceptual perspectives. In the case of empirical concepts there exists a well known situation in which conflict arises on the basis of two positive accounts being given of one state of affairs. And such conflict, in the final analysis, is not a logical or formal one. It is the conflict that arises in the simultaneous application of different colour concepts. In this regard Bertrand Russell has said: "'this is red' and 'this is blue' are incompatible. The incompatibility is not logical. Red and blue are no more *logically* incompatible than red and round. Nor is the incompatibility a generalisation from experience."[8] It has been suggested by D. J. O'Connor that the lack of formal conflict here is a result of the fact that "'red' and 'blue' and terms like them ... can be defined only ostensively." And it has been thought by a number of philosophers that such incompatibility is restricted only to a small range of empirical concepts which are dependent on ostension for their meaning, including perhaps certain shape concepts, such as 'round' and 'square.' Other empirical concepts, they believe, can be analysed so as to reveal an underlying contradiction, which may be taken to account for their conflicting.

But if we take as the criterion for a concept's being empirical that its definition ultimately rest on ostension, then, as has been argued by Lionel Kenner, "the difficulty which is seen with colour words applies, in the end, to all empirical concepts."[9] Thus, though the considerations of the above authors have not led them to the distinction made in the present study between the application of an empirical concept and the claim that the concept applies, Kenner's argument nevertheless suggests that even in the latter case the nature of the incompatibility between such concepts would not be logical or formal. And on the present view, where we are not concerned with the

---

[8] Russell (1940), p. 100; next quote, O'Connor (1955), p. 112.

[9] Kenner (1965), p. 151. Kenner however takes this to imply the *triviality* of the problem with colour concepts, while it is here seen as suggesting its *generality*.

making of statements but only with the application of concepts, we should say that in their application to the whole of one and the same thing, distinct empirical concepts might well be perspectivally incompatible.

Formal or Syntactical Contradiction vs. Contradiction Proper vs. 'Contrariety'

In order to clarify the view taken here it may prove helpful to distinguish among certain sorts of incompatibility which are candidates for characterising the nature of the conflict occurring in the case of successive scientific theories. To this end then we might begin with the notion of formal, syntactical, or logical contradiction. Such a 'contradiction' manifests itself in the syntax or form of certain linguistic expressions – more particularly, in the form of sentences; and it involves the conjunction of a sentence with its negation. Thus we should say that the (molecular) sentence "This is red (all over), and it is not the case that this is red (all over)" is a formal contradiction. The essence of this contradiction can be captured by formulating the sentence in the sentential calculus as: $A \wedge \neg A$, or in the predicate calculus as: $Pa \wedge \neg Pa$. Thus the nature of the conflict between competing scientific theories as conceived on the Deductive Model is of this sort.

Formal or syntactical contradiction may be distinguished however from contradiction proper in that where the former is a relation between sentences and involves negation, we may say that the latter is a relation between statements or, perhaps better, between *propositions*, and involves *denial*. Thus, to take an extreme example, if a person were to say "This is red (all over) and it is not the case that this is red (all over)," and meant a different thing by "this" or "red" on each of their uses, while we would have a syntactical contradiction, we would not have a contradiction proper. (This is essentially the problem of meaning variance.)[10] While sentences may be used to express propositions, the form of the sentence is not itself a guarantee that the proposition it is being used to express is or is not a proper contradiction.

---

[10] For a (very) early discussion of this problem which is quite in keeping with the present view, see Campbell (1920), pp. 51ff.

A third notion of relevance to the present case will here be called 'contrariety,' and may be exemplified by a statement such as 'This is both red (all over) and blue (all over).' Contrariety thus differs from contradiction proper in that its faithful linguistic representation does not, prima facie, evince a syntactical contradiction, and in that it necessarily involves the employment of 'positive' alternatives, and thus cannot be simply a case of affirmation and denial. Thus, for example, the formulation of the above statement in the predicate calculus would give: $Pa \wedge Qa$; and this syntactical representation fails to capture the conflict involved in the statement it represents. Furthermore, as is evidenced by this formulation, where it should always be the case that one of two contradicting statements be true, both of two contrary statements may be false.

The notion of perspectival incompatibility thus comes closest to what has here been called 'contrariety,' but differs from it in the essential respect that it does not involve statements or propositions having a truth-value, but rather the application of concepts. Thus where the propositions 'This is red (all over)' and 'This is blue (all over)' evince 'contrariety,' the attempt simultaneously to apply the concepts 'red' and 'blue' to the whole of one thing constitutes an instance of perspectival incompatibility.

Conflicting Perspectives Suggesting the Same Results

It may be emphasised that perspectival incompatibility arises in such cases independently of there being any sort of anomaly favouring one perspective at the expense of the other, and that two concepts being perspectivally incompatible does not in itself imply that either of the concepts taken alone is not applicable to the state of affairs in question. At the level of simple empirical concepts this may be made clear by considering the Necker Cube.[11] The drawing cannot be seen as two sorts of cube at the same time – in this way the two cube concepts are perspectivally incompatible; nevertheless, unlike the case involving the duck-rabbit, the drawing is not itself more in keeping with one of the concepts than with the other.

In considering actual science a situation similar to this arises, for example, in the case of the Copernican and Tychonic (note: not

---

[11] Cf. e.g. Wittgenstein (1921), § 5.5423.

Ptolemaic) systems of astronomy.[12] Each system conflicts with the other in constituting its own conceptual perspective on the motion of the planets, but both systems suggest their having the same motions relative to one another and to the sun. Thus, in application to the solar system, we might say that these systems are not incompatible at the level of observation, while they are incompatible at the level of theory. Another example of this sort is provided by the caloric and dynamical theories of heat, as described by Kuhn: "In their abstract structures and in the conceptual entities they presuppose, these two theories are quite different and, in fact, incompatible. But, during the years when the two vied for the allegiance of the scientific community, the theoretical predictions that could be derived from them were very nearly the same."[13]

The existence of these sorts of cases in science leads directly to a questioning of the Popperian conception of theory conflict, for on Popper's view theory clash is to *depend* on the theories concerned giving conflicting results, such that if both of two theories suggest the same results they ought not be considered as being incompatible. Popper says: "in order that a new theory should constitute a discovery or a step forward it should conflict with its predecessor; that is to say, it should lead to at least some conflicting results. But this means, from a logical point of view, that it should contradict its predecessor." This picture is of course quite in keeping with Popper's view as based on the Deductive Model, but it means that he is unable to account for the conflict that may arise in those cases where the theories concerned do not differ with regard to their predictions. But this in turn suggests that the situation is not simply that Popper's conception of theory conflict is inapplicable in a few odd cases, but rather that even in those cases where distinct theories do suggest incompatible results, this difference between them is not essential to

---

[12] Cf. Kuhn (1957), pp. 201ff. Cf. also p. 96, n. 9, below. Other examples include the Machian vs. Newtonian conceptions of space (below, p. 200), the eccentric vs. epicycle hypotheses regarding the motions of the planets (see Heath, 1913, p. 266), and a thought-experiment involving perfectly circular orbits about the sun, in the one case their being conceived of as due to a gravitational attraction, and in the other as due to circular motion being natural (*à la* Aristotle). For a discussion relevant to the present one, see Hanson (1966).

[13] Kuhn (1961), pp. 176–177; next quote, Popper (1975), pp. 82–83.

their conflicting; and if this is so, then Popper's notion of theory
clash does not capture the essence of theory conflict in *any* case in
which it occurs.

Meaning Variance

The difficulty of applying the Popperian concept of theory conflict to
the case of actual theories has been treated earlier with regard to the
problem of meaning variance. Here we will look at this problem somewhat more closely, and show how it can be overcome on the present
conception. An early formulation of the problem is given by Dudley
Shapere, though he sees it as being a problem for Feyerabend's view
rather than for those philosophies of science based on the Deductive
Model. With regard to Feyerabend, Shapere asks:

How is it possible to reject *both* the consistency condition *and* the condition
of meaning invariance? For in order for two sentences to contradict one
another (to be inconsistent with one another), one must be the denial of the
other; and this is to say that what is denied by the one must be what the
other asserts; and this in turn is to say that the theories must have some
common meaning.[14]

If we approach this problem first on the level of basic epistemology, and take the meaning of a (descriptive) term to be the (empirical)
concept it is being used to express, earlier considerations suggest that
in certain cases conflict can occur precisely *because* there is a change
in the meaning of such a term.[15] If one person says that something is
red, and is using the term "red" as it is normally used, while another
person says that the same thing is red, but in saying so actually
means 'blue' by the term "red," then we have an instance of the sort
of conflict earlier called 'contrariety.'

Or, to take an example from the history of language, we might
consider the English term "nice." A few centuries ago this term had a
meaning 'wanton' or 'lascivious,' and it has now come to mean

---

[14] Shapere (1966), p. 57. See also e.g. Achinstein (1964), p. 499, and Giedymin
(1970), p. 265. Note also Stegmüller's remark that, as criticisms of the views of
Kuhn and Feyerabend, "*Almost all* of Shapere's arguments are based on the statement view and, for the most part, lose their force with its rejection." (1973), p. 261.
[15] M. Hesse touches on this point and provides an example relevant to it in her
(1963), pp. 102–103.

'kind, considerate, pleasant to others.'[16] If we grant that a person cannot be lascivious and at the same time kind and considerate, we can see how the change in the meaning of "nice" would *account* for the conflict involved in saying of a person that he is nice, meaning 'lascivious,' and that he is nice, meaning 'kind.'

This same reasoning can be applied to the much discussed issue regarding the meaning of the term "mass" in the context of Newton's and Einstein's gravitational theories respectively. As used in the context of Newton's theory, implicit in its meaning is the idea that the mass of a body is independent of its velocity. The same term, employed in the context of Einstein's theory, implies on the contrary that the mass of a body increases with its velocity. In this way then we might say that a change in the meaning of the term "mass" would account for the conflict involved in saying of one and the same moving body that it has some one particular mass both in Newton's sense and in Einstein's.

The above examples are intended mainly to suggest that different statements may be inconsistent even when the descriptive terms common to them have different meanings (senses). As has been mentioned above however, on the present conception theory conflict is not viewed as being based on the inter-theoretical relations among the meanings of individual terms, but is seen rather as resulting from the attempt to apply the whole of certain distinct theories to the same state of affairs. Thus while the present view allows for the meaning variance of individual terms, and while the differences in the particular natures of the respective theories might in certain cases be seen as resulting from the sum of these sorts of conceptual differences, it is the theories as wholes that are here seen to conflict. In other words, on the present view incommensurability between theories involves a shift of conceptual perspective – a shift which may or may not be accompanied by a meaning change of individual terms.

The preceding discussion thus provides a means by which Shapere's criticism of Feyerabend with regard to meaning variance and incompatibility can be overcome. In spite of this however, unless Feyerabend is willing to allow that two incommensurable theories can both pertain to one and the same state of affairs, his view of

---

[16] Ullman (1962), p. 234.

incommensurability is still open to charges of relativism which the present conception succeeds in avoiding.

5. INCOMMENSURABLE THEORIES HAVING THE SAME INTENDED DOMAIN

Throughout his writings in the philosophy of science Feyerabend has maintained a form of realism – 'epistemological realism' – which suggests that in accepting a scientific theory one is to take the theoretical constructs appearing in the theory as having real counterparts with the properties predicated of them by the constructs. Thus, on Feyerabend's view, if two theories have in their respective bases essentially different constructs, then the entities in the world to which these theories apply, if they apply at all, are themselves to have essentially different properties. And, so the reasoning goes, seeing as such entities are referred to, if not solely then primarily, by those constructs in terms of which they are characterised in such theories, then theories based on essentially different constructs ought to be taken as referring to essentially different things. We might in fact say that on most occasions where Feyerabend speaks of theories being incommensurable, he has in mind that they differ in this sort of way. Thus we read, for example, in his *Science in a Free Society*:

... All we need to do is to point out how often the world changed because of a change in basic theory. If the theories are commensurable, then no problem arises – we simply have an addition to knowledge. It is different with incommensurable theories. For we certainly cannot assume that two incommensurable theories deal with one and the same objective state of affairs (to make the assumption we would have to assume that both at least refer to the same objective situation. But how can we assert that 'they both' refer to the same situation when 'they both' never make sense together? Besides, statements about what does and what does not refer can be checked only if the things referred to are described properly, but then our problem arises again with renewed force.) Hence, unless we want to assume that they deal with nothing at all we must admit that they deal with different worlds and that the change (from one world to another) has been brought about by a switch from one theory to another.[17]

---

[17] Feyerabend (1978), p. 70.

With regard to this line of reasoning we note that even on its weaker reading (in which the world does not change) it creates difficulties when it comes to explaining the experiments in actual science (e.g. Michelson-Morley) which have been considered crucial with respect to the viability of the sorts of theories Feyerabend takes to be incommensurable. If incommensurable theories concern different worlds – are about different things – then how are we to conceive of them as competing, or winning or losing in their competition? Feyerabend's 'pragmatic theory of observation,' for example, does not answer this question, for it suggests that each theory has its own criteria of acceptability.[18]

On the present view however this relativistic consequence is avoided, for here we should say that when a scientist moves from one of two incommensurable theories to the other, he can still very well *intend* that both theories apply to the same states of affairs, even if they characterise those states of affairs in essentially different ways. Furthermore, on the basis of certain criteria (such as the performance of the same sorts of operations), we, as onlookers, can often judge that the scientist is treating of one and the same aspect of reality in his respective applications of the two incommensurable theories.

In fact, in *Against Method*, Feyerabend has himself discussed the notion of incommensurability in such a way as can be naturally extended to this view of the situation. In keeping with the present conception as based on the Gestalt Model, as mentioned in Chapter 7 above he has suggested that gestalt-switch figures provide instances of incommensurability – and in such cases it is clear that the incommensurable concepts concerned are both being referred to the same thing, namely, the gestalt figure itself. And, in another place, Feyerabend expresses a view similar to the present one in saying that:

When we go from classical physics to relativity, what remains the same are the objects. The objects are what they are, only we think different things about them. To a large extent, our operations for getting certain numbers

---

[18] For a discussion of Feyerabend's 'pragmatic theory,' with relevant references, see Shapere (1966), pp. 59ff. See also Feyerabend's discussion of crucial experiments in e.g. his (1970), pp. 226ff.

remain the same. The functors which we use and much of the syntactical apparatus may also remain the same. What is different are all the concepts connected with the functors.[19]

Thus, in keeping with this line of thought, it is here suggested that even on a realist conception incommensurable theories might well pertain to the same states of affairs, or have the same *intended domain*. Of course there exist theories which are intended to treat of different things, and, if one wanted, one might say of such theories that they are incommensurable. But they would not be incommensurable in the interesting sense treated here, where the theories concerned conflict with one another.

Speaking more generally, we should here say that differing conceptual perspectives can have the same *reference*.[20] The use of this term is to suggest the idea that a conceptual perspective, in being an applied concept or system of concepts, in a sense 'points' in a certain direction. And the direction in which it 'points' is in turn dependent upon the *intention* of the person applying the concept or conceptual framework, and not on the concept itself. Thus, as has been suggested in the previous chapter, the 'duck' and 'rabbit' concepts of the Gestalt Model might each be intended to apply to a different figure, and consequently be said to be given different references. It may also be noted that where a concept or system of concepts is *given* a certain reference, a conceptual perspective, in consisting of an *applied* concept or framework, *has* a particular reference. And it is only when two perspectives have the same reference that they conflict.

Furthermore, it may be pointed out that in order for such conflict to arise it is not the case, on the present view, that that towards which the perspectives are directed need actually exist. (The duck-rabbit figure might be a mirage – but even if it were, the 'duck' and 'rabbit' concepts could not be applied to it simultaneously.) The conflict is in this way dependent on the intention(s) of the person(s) employing the concepts, and not on the existence of that to which it may be believed that they are being applied. Thus in the case of scientific theories for example, we might have two different theories which

---

[19] Hanson et al. (1970), p. 247.
[20] In Dilworth (1978) the term "intention" was used to cover both of the notions 'intention' and 'reference' treated in the present study.

conflict with one another as regards the nature of certain subatomic particles – particles which further research reveals not to exist at all.

6. SYSTEMS INVOLVING THE SAME CATEGORIES

In the present conception the notion of commonality of reference affords a way of conceiving how two radically different theories might well be about the same realm of phenomena, or have the same intended domain. In order for them to conflict though, they must also involve the same *categories*.

The individual predicates 'red' and 'blue,' as versus the predicates 'red' and 'round,' are of the same category, 'colour,' just as in the Gestalt Model the concepts 'picture-duck' and 'picture-rabbit' may be thought to be of the same category 'shape.' Scientific theories however, in that each constitutes a more complex sort of conceptual perspective, rather than simply being of just one category, may each be thought to involve a number of categories. And, in considering more than one theory, the categories they involve may or may not be the same.

The distinction between the concept of category and that of class or set helps mark a difference between the present account and those of the empiricists and Popperians. Allowing that predicates can be grouped according to their category, and that things may be grouped into classes depending on the applicability of certain predicates, on the empiricist and Popperian views scientific laws and theories may most simply be seen as claiming that one class or set of things is a subset of another, that is, that the class of all things x to which a particular predicate F is applicable is a subset of the class of all things y to which a particular predicate G is applicable. And conflict is thought to arise when a further claim is made to the effect that there exist things in class F which are not in class G.[21]

Now this approach gives rise to basic ontological, epistemological, and logical problems, a detailed discussion of which lies beyond the scope of the present study. These problems include, for example, how we are to determine what is to constitute a thing in our universe,

---

[21] For a similar description of this approach, and a discussion of it, see Wartofsky (1968), pp. 277ff.

and what we are to do in cases where it is debatable whether some such thing really does or does not possess some particular property, as well as what steps are to be taken so as to avoid the possibility of formulating the classical logical paradoxes which arise from the un-restricted use of the principle of abstraction, whereby a class or set is defined on the basis of its elements having a certain property.

On the present account however, in keeping with the view of Evandro Agazzi, the objects of which science treats are given by certain *operations*; or, in the case of physics, they consist more par-ticularly in the *data* resulting from certain *measurements*. Thus we follow Agazzi where, for example, he suggests that we

[consider] what the different sciences do in order to treat 'things' from their 'viewpoint': they submit them to certain specific manipulations of an *oper-ational* character, which put the scientist in the position of answering cer-tain specific questions he can formulate about these things. Such operational procedures may be the use of a ruler, of a balance, of a dynamometer, in order to establish some physical characteristics of the 'thing' like its length, its weight or the strength of some force exerted on it; they may be the em-ployment of some reagents to determine its chemical composition, etc.[22]

In the same vein then, we should here say that evidence suggesting the particular categories with which a certain theory is involved may be obtained via a consideration of the sorts of operations or measure-ments used in applying or testing the theory. Thus two distinct theo-ries' pertaining to the same measurements suggests that they involve the same categories.[23] The above may be brought out by considering the application of colour concepts. It may be wondered, for example, whether some particular thing is more red than orange, or vice versa, i.e., whether one of the predicates 'red' or 'orange' is the better applicable. The scientific approach would be to take a reading on the

---

[22] Agazzi (1976), p. 148. Note that Agazzi's 'viewpoints' are here more closely aligned with different sciences than with different theories. For a development of this view see the rest of Agazzi (1976), as well as Agazzi (1977a).

[23] Kuhn's view of incommensurable theories thus differs from the present one to the extent that he suggests that such theories need pertain to different data; in this regard cf. Kuhn (1962), p. 126, and (1974), p. 473n.

thing in question using a spectrometer.[24] And the fact that the same operation is of relevance to the applicability of both predicates suggests that both are of the same category (viz., 'colour').

The distinction between theories which are perspectively incompatible and those which are not may in turn be made clearer through a consideration of the following simple schema.

Category A 'colour'	Category B 'shape'	Category C ...
1. 'red'	1. 'round'	1. ...
2. 'blue'	2. 'square'	
3. ...	3. ...	

Table 1

We may say that the determination of the applicability of the predicates in the category 'colour' involves the use of a spectrometer, and the applicability of those in the category 'shape' is determined by the use of a ruler and compasses. That different operations should be performed implies that the predicates are in fact of different categories. And the attempt to apply, for example, the predicates 'red' and 'round' to the whole of one thing would not result in conflict. In being of different categories, 'red' and 'round,' as applied predicates or concepts, would not be perspectively incompatible.

---

[24] Cf. a slightly different example given in Allen & Maxwell (1939), p. 5: "We can better understand the difficulties of the earlier investigators if we consider the question of *measuring* colour.

"Colour we generally regard as a quality. It is, however, possible to select a scale in which a particular colour, blue for example, is graduated in depth from very pale to very dark. We could go further and, by making a *mental estimate,* attach a series of numbers to the various samples so that we could speak of any given sample of blue as having so many 'degrees of blue.' This is not quite the same process as a *physical measurement,* but we might take a further step and try to find a correlation between our mental scale and a physical scale derived, for example, from the amount of dye used in preparing a specimen or from the relative proportions of the blue and white sectors in a rotating disc."

With regard to the relation between operations and scientific objectivity in the context of colour concepts, see Agazzi (1978), pp. 100ff.

A situation like this can occur in the case of scientific theories. One theory might be concerned with the motion of certain bodies, while another might consider those bodies solely from the point of view of the electromagnetic radiation being emitted from them. To the extent that the operations performed in the application of the two theories should differ, the theories would not be perspectively incompatible.

Agazzi has made a point similar to the above, and has extended this way of thinking to provide a criterion for demarcating different sciences:

[A]ny science delineates and determines its object sphere by means of a finite series of basic predicates. Such predicates are always defined in terms of operative criteria and they serve to provide this science with its objects. For instance, something is to be regarded as an object of mechanics if when speaking about it we use as basic concepts only concepts of mechanics (such as mass, space, time, force). ... Thus, we could say that any science determines through its basic predicates its own *whole* scope so that everything outside this scope is of no interest to it and lies outside its frame of reference. For example, electromagnetic phenomena are not part of mechanics because they cannot be dealt with in terms of the basic mechanical concepts.[25]

While two distinct theories would not be perspectively incompatible if they involved the application of predicates from different categories, they could be incompatible if they respectively involved the application of predicates from within the same category (to the same intended domain). Where, for example, the predicates 'red' and 'round' can both be applied to the same thing, the predicates 'red' and 'blue' (or 'round' and 'square') cannot. The case is similar with different scientific theories: if they are related to the same operations, and consequently concern the same categories, then, other things being equal, they would be perspectively incompatible, independently of whether they suggested the same or different results of those operations.

---

[25] Agazzi (1977b), p. 166. This quotation also has direct relevance to the tables appearing at the beginning of the next chapter.

7. RELATIVE ACCEPTABILITY: ACCURACY, SCOPE AND SIMPLICITY

The presentation of the Gestalt Model in the previous chapter reveals three sorts of factor that may be taken into account in considering the relative acceptability of conflicting conceptual perspectives: accuracy, scope and simplicity. Here, competing scientific theories, in constituting conceptual perspectives, are also to be thought of as comparable with regard to these factors, and on the basis of such a comparison may be judged as to their relative acceptability; and we can say that *progress* has been made when a theory is adopted which is, in terms of these factors, relatively more acceptable than its predecessor.

Accuracy

The example involving colour concepts used above to indicate the relevance of categories to the present view also shows the way in which one of two competing theories may be judged on the basis of measurements to be more accurate than the other. Thus, the readings obtained on a spectrometer might show one of the predicates 'red' or 'orange' to be the better applicable to some particular thing. And the same may be the case with regard to scientific theories: measurements which are relevant to the applicability of two competing theories may show one of them to be the more accurate – even in the case where neither theory suggests exactly those results obtained. Thus on the present view, while such theories would be *incommensurable* in that the move from one to the other involves a shift of conceptual perspective, they might nevertheless be considered *commensurable* in the sense that, and to the extent that, the same operations are employed to determine their respective applicability. And, where competing theories need not suggest different results in order to conflict (consider the example of the respective astronomical systems of Tycho and Copernicus), they must do so in order for one of them to be judged as being more accurate than the other.

But the theories' suggesting different results is not in itself sufficient to compare them in this regard. The results they respectively suggest must also differ sufficiently so as to lie outside of the range of mensural error of the instruments employed in the comparison.

Thus the testing of certain theories as to their relative accuracy might have to await the development of sufficiently sensitive measuring instruments; and the failure to develop such instruments would preclude this sort of comparison.

Also, competing theories may for the most part suggest respective results which do not lie outside of this range of mensural error, but which are measurably different only when the theories are applied to a particular special case. In such a case then we can employ measuring instruments so as to stage a crucial experiment, the results of which may be taken to suggest one of the theories as being the more accurate.

Scope

An independent factor which may play a role in a debate concerning the relative acceptability of incommensurable theories is the scope or generality of each of them. Thus, in the simplest case, where two such theories are equally accurate in application to the intended domain they have in common, and one of them is also more or less applicable to states of affairs to which the other does not apply at all, then, other things being equal, it would have the greater scope of the two and would for this reason be the more acceptable. Of course there is nothing to prevent more complicated situations from arising in which, for example, one theory has a wider scope than the other, but the other is the more accurate in application to the domain to which both more or less apply. And it is clear that in such a case the debate over which of the two theories is all in all the more acceptable cannot be so easily resolved; and both theories might be retained, each to be used in those particular cases where the sort of superiority it evinces can come to the fore.

Simplicity

Where the notions of accuracy and scope indicate empirical factors which may play a role in the possible argumentation concerning the relative acceptability of competing theories, the notion of *simplicity* indicates a potential factor which is of a more rational nature. As is suggested by the Gestalt Model, the relative simplicity of a scientific theory may be likened to the ease or naturalness with which a gestalt diagram can be seen in one aspect as compared with another. Following the model further, we may say that the simpler of two theories is the one involving the fewer ad hoc modifications in its attempt to explain the relevant empirical data, i.e., the one which is the less *sophisticated*.

An example of relevance to this notion of simplicity is provided by a consideration of the astronomical systems of Ptolemy and Copernicus. This example is particularly suitable here, for the two systems have the same scope, and may be taken to be equally accurate. But in terms of its basic concepts Copernicus' system can give a good qualitative explanation of such states of affairs as the retrograde motion of the planets, the velocities of the planets in their orbits, the fact that Mercury and Venus never stray far from the sun, and so on. The basic functioning of these sorts of phenomena is *integral* to Copernicus' system, and the irregularities they evince are simplified and shown only to be apparent. On Ptolemy's system however, in its simplest form, these phenomena find no place at all, and it is only through ad hoc modifications that they can be accommodated. And where on Copernicus' system their apparent irregularity is shown to be illusory, on Ptolemy's it is represented by an irregularity in the system itself. Thus the planets do not only appear to loop back in their orbits, they are conceived as actually doing so; each planet does not only appear to move with an inconstant angular velocity as relative to the centre of its orbit, but is thought actually to behave in this way; and the handling of the fact that Mercury and Venus are never seen far from the sun is, on the Ptolemaic system, entirely ad hoc, and consequently does not explain this phenomenon at all.[26]

Now this is not to say that the full-blown version of Copernicus' system is free from sophistications, for as we realise today the orbits of the planets are more nearly elliptical than circular, with the sun standing at one of the foci of the ellipse described by each planet, and Copernicus had to modify his theory when faced with factual discrepancies. (On top of this it seems that a number of the modifications he made to his original scheme were due to his working from erroneous data.) Thus even Copernicus' system was submitted to

---

[26] As regards this last point see Kuhn (1957), pp. 172–173. For the other points see the same work, esp. Chh. 2 and 5.

ad hoc modifications; and the resulting scheme might well look as complicated as Ptolemy's. But the above considerations nevertheless suggest Copernicus' system to be the more coherent, and to be essentially simpler than Ptolemy's in the sense intended here.[27] We might say that it provides a conception of the motions of the planets which is easier to grasp 'in a single gestalt.'

Thus we see that on the present view not only is theory conflict explainable as perspectival incompatibility, but the problem of meaning variance is overcome, and a positive conception is given of the sorts of factors involved in determining whether one scientific theory should constitute a progression beyond another. In the next chapter the present view will be further developed in the context of an example taken from science, and it will there be seen to handle a number of other notions of philosophical interest as well, including those of idealisation and theoretical terms, and it will also be seen to provide a more realistic conception of the nature of scientific theories than that afforded on the basis of the Deductive Model.

---

[27] Thus the notion of simplicity discussed here is a relative one, and the present explanation should answer Lakatos' and Feyerabend's questions as regards how Copernicus' theory is simpler than Ptolemy's: see Lakatos (1970), p. 117&n.; and Feyerabend (1978), p. 47.

CHAPTER 10

DEVELOPMENT OF THE PERSPECTIVIST CONCEPTION IN THE CONTEXT OF THE KINETIC THEORY OF GASES

1. PARAMETERS AS QUANTIFIED CATEGORIES

The conception of science and scientific progress presented in the previous chapter may be further explicated with the help of an example taken from the physics of gases. Though the presentation of this example will for the most part follow the actual development of gas theory, it is not intended to constitute the basis of an historical analysis, but to be a coherent reconstruction capturing the essence of the conceptual moves in this development. As a first step in the presentation of the example, the sort of schematisation provided by Table 1 (p. 83) is here given a more definite form as a table of particular *parameters*, or quantified categories; and the general remarks made in the context of Table 1 should also be applicable here.

	Parameter A mass	Parameter B length	Parameter C time	Parameter D temperature
Value scale	$\mathbb{Q}$	$\mathbb{Q}$	$\mathbb{Q}$	$\mathbb{Q}$
Unit	kilogram	metre	second	degree Kelvin
Measuring instrument	balance	metre stick	clock	thermometer

Table 2

In comparison with Table 1, predicates or concepts falling under a certain category are here rational number *values* of a parameter, expressed in terms of the *unit* specific to the parameter. And with each

parameter is associated a measuring instrument, the appropriate use
of which should allow the determination of the value of the parameter in a given empirical situation.[1]

From the parameters appearing in Table 2 we can obtain further
parameters which are of direct relevance to the kinetic theory of
gases:

	Parameter A´ volume	Parameter B´ pressure	Parameter D temperature
Value scale	Q	Q	Q
Unit	cubic metre	newton per square metre	degree Kelvin
Measuring instrument	metre stick	manometer	thermometer

Table 3

The parameters volume and pressure appearing in Table 3 are each
derived from parameters appearing in Table 2. Volume is derived
from length, and pressure is derived from mass, length, and time via
the intermediate parameter *force*; and both volume and pressure are
assumed to take rational number values.[2] The unit *newton* is that of
the parameter force, and is defined as kilogram times metre per
second squared. The parameter temperature is carried over directly
from Table 2.

---

[1] The standardisation required in order for this and the following table to be applicable to a development spanning more than two hundred years has been facilitated by
the employment of notions of contemporary science: e.g. those of *newton* and
degree Kelvin appearing in Table 3 below. Also, following standard notation, quotation marks are not being used in referring to individual parameters. Nevertheless,
parameters, as quantified categories, are not here conceived as existing in the world,
but rather as being abstractions we employ in our attempts to understand it. Cf.
p. 62, n. 17, above.
[2] For a description in which parameters can take real number values see the text to
n. 10, p. 98, below.

2. BOYLE'S LAW

In terms of Table 3, *Boyle's law* (1661) states that the (value of the) pressure times the (value of the) volume of a given gas, as measured by a manometer and metre stick respectively, is constant, given constant temperature, as measured by a thermometer.[3] This may be abbreviated to:

(28) $pV = R$ (given constant temperature).

As presented here, Boyle's law constitutes what physicists would call an *empirical* or *experimental* law. While it does express a regularity of nature, the law itself is based solely on results obtained using instruments in actual measurements, and is not here related to theory. It does not succeed in explaining why, given constant temperature, the volume of a given mass of gas should be inversely proportional to its pressure; it tells us only that this is the case.

In application to real gases Boyle's law does not hold exactly, giving particularly divergent results from those obtained by measurement in the case of gases at high pressure and low temperature, i.e., when the gas in question approaches that critical point at which it begins to condense into a liquid. A gas to which Boyle's law applies exactly may thus be defined as an *ideal gas*. A formula subsuming Boyle's law, involving the further realisation that the product pV varies proportionally to the temperature (T), constitutes the equation of state for an ideal gas:

(29) $\dfrac{pV}{T} = R.$

This same equation, when applied to *real* systems, is called the *general gas law*, and serves to express the relation expected to obtain among the values of all three of the parameters volume, pressure and temperature.

---

[3] In its original formulation, Boyle's law did not involve the parameter temperature. In subsequent developments however it was realised that the applicability of the law requires that temperature be held constant.

3. THE IDEAL GAS MODEL

A model for an ideal gas has been developed in the context of the kinetic theory of matter (Bernoulli, 1738).[4] According to the model as understood today, an ideal gas consists of molecules in motion, and has the following further properties:

1. The volume of all the molecules taken together is negligible in comparison with the volume of the container occupied by the gas;
2. there are no forces acting on or between the molecules except in the case of collision;
3. when colliding with each other or with the walls of the container the molecules act as completely elastic spheres;
4. the time a molecule takes in colliding is negligible as compared with the time between its collisions;
5. the motion of the molecules is completely random.

Boyle's law can be derived from the model for an ideal gas. The derivation involves assuming, among other things, that a particular sample of the gas contains some definite number of molecules each with the same mass and all of which share a certain average velocity.[5] We can also derive from the model the experimental law of Gay-Lussac (1808) concerning the relation between the volumes of interacting gases in the case where chemical changes occur. When so derived, such experimental laws are said to be *explained* by the model (or by the theory in which the model plays a central role). We can also say that the model explains, for example, the pressure of a gas, as manifest in manometer readings, as being the result of molecules bombarding the walls of the container.

As mentioned above, Boyle's law, or the general gas law, fails to give values close to those obtained via measurement at high pressures and low temperatures. More particularly, it gives no hint that, when a gas at a temperature below a certain critical point is highly

---

[4] Bernoulli is generally recognised as being the first to suggest a model of the sort which is today called the ideal gas model. It may be noted however that in Bernoulli's model there is to be an infinite number of molecules: cf. Partington (1961), p. 477.

[5] This is the approach taken e.g. in Barton (1933), pp. 197–201. It seems however that the derivation should also be possible even assuming an infinite number of molecules, as in the case of Bernoulli's model: cf. Partington (1961), p. 477.

compressed, it will undergo a significant decrease in volume and become a liquid.

One way to account for this decrease in volume on the kinetic theory of gases is to assume that there is an attractive force operating between the molecules which becomes effective when the molecules are in relatively close proximity to one another (i.e. when the gas is under high pressure), thus sharply repressing their motion at a particular critical point. Also, it may be noted that while it is understandable that at low pressures the total volume of the molecules of a gas may be considered negligible as compared with the volume of the vessel containing the gas, as the gas is compressed this assumption becomes moot.

But both of these notions – that of an attractive force operating between the molecules and that of their having a non-negligible volume – run counter to the ideal gas model. Thus if these factors are to be taken into account in attempting to explain the behaviour of gases under high pressure, as well as the phenomenon of change of state, a new model must be devised.

4. VAN DER WAALS' LAW

Still in accordance with the kinetic theory of matter, van der Waals (1873) developed a model capable of being used to explain the behaviour of gases under high pressure, as well as their changes to the liquid state. In the model inter-molecular attractive forces are present, and molecules are assigned a definite volume. The equation of state of van der Waals' model may be presented in the following form:

$$(30) \quad \left(p + \frac{a}{V^2}\right)(V - b) = RT,$$

where a and b are constants characteristic of the gas in question. More particularly, the expression "a/V^2" represents the decrease in pressure (as expected on the equation of state of the ideal gas model) exerted by a substance as a result of the effect of inter-molecular attractive forces; and the term "b" represents the effected decrease in the molecule-free volume of the container as a result of the volume of the molecules. For $a = b = 0$, (30) reduces to the equation of state for an ideal gas.

As an expression of van der Waals' law, (30) is thus to have application not only to substances in their purely gaseous state, but also to substances as they undergo a transition from their gaseous to their liquid form.

If the pressure, volume, and temperature are expressed as fractions of the critical (i.e. 'change of state') pressure, volume, and temperature, which are constants, we obtain van der Waals' *reduced equation of state*:[6]

$$(31) \quad \left(\mathscr{P} + \frac{3}{\mathscr{V}^2} \right) \left(\mathscr{V} - \frac{1}{3} \right) = \frac{8}{3} \mathscr{T}.$$

This equation does not contain the constants a and b, which differ for different gases, and so is a generalised expression which applies to all gases to which the individual equations having the form of (30) apply.

While van der Waals' equation does not succeed in giving exactly correct results, it nevertheless constitutes an advance over the general gas law in that it finds application to the phenomenon of change of state, in most cases providing results of at least the correct order of magnitude.

5. EXPLICATION OF THE PERSPECTIVIST CONCEPTION IN THE CONTEXT OF THE EXAMPLE

In the previous chapter the basic notions of the present conception were presented and discussed against the background of the Gestalt Model, and involved the employment mainly of simple empirical predicates such as 'red' and 'round.' The central notions treated there were those of conceptual perspective, logical simultaneity, perspectival incompatibility, intention, category, and relative acceptability. In what follows all of these notions will be treated once again, but this time with regard to the example of the present chapter, thus delineating more clearly their relation to actual science.

---

[6] For details see e.g. Mitton (1939), pp. 179–182.

Of particular interest to the present view is the role played by *models* in the above example. Models will here take over the position occupied by *concepts* and *predicates* in Chapter 9. Thus while models and predicates, in application, may each be thought to constitute a *conceptual perspective*, scientific theories are conceived more particularly to be applied models.[7] It may be noted that theories are not here being identified with models: an unapplied model is not a theory; and while every scientific theory is taken to be some sort of applied model, not every applied model need be a scientific theory.

On the present view, that to which a model is applied is determined by the *intention* of the individual applying it; and it may be applied either to real or to imaginary cases: e.g., the ideal gas model may be applied either to real or to ideal gases. We should thus say that the *reference* given the model, or the *intended domain* of the theory, differs in the two cases. More generally though, the intended domain of a scientific theory is to be thought of as encompassing all of the empirical states of affairs to which it is intended that its model be applied. Thus we see that the reference given van der Waals' model is broader than that given the ideal gas model, in that it is intended to be applied not only to substances in their purely gaseous form, but also to such substances when they undergo a change of state.

If the parameters *a* and *b* of the van der Waals equation of state are given the value zero, then van der Waals' equation reduces to the equation of state for an ideal gas. As a development of the present view, we can more generally say that the notion of *reduction* is to apply to those cases where a more complex equation gives results identical to those of a simpler equation when certain of the parameters occurring in the former are given limiting values (often zero). When this is so, we should thus say that the more complex equation *reduces to* the simpler one, or, we may say that the simpler equation, or the model from which it is derived, constitutes a *limiting case* of

---

[7] In this regard cf. W. A. Wallace (1974), p. 263: "[I]n many situations where a novel modeling technique is employed to gain understanding of a phenomenon, a new way of looking at things is involved and a type of Gestalt switch may take place. In this sense Kuhn is quite correct in seeing scientific revolutions as involving such switches and changed viewpoints. In fact, his paradigm shifts can very frequently be seen as modeling shifts."

the more complex equation or model respectively. Now, it is to be noted that in the physical interpretation of van der Waals' equation the parameters a and b are not to be given the value zero, but are each rather to assume some constant positive value depending on the substance under investigation. Thus the reduction here is a purely formal one, and it does not obtain in any instance where van der Waals' equation is being treated as the expression of a law of nature, or as the equation of state of van der Waals' model.[8]

The fact that the parameters a and b in van der Waals' equation of state always receive some positive value, i.e., the fact that in his model inter-molecular attractive forces are always assumed to be present and the molecules are always taken to have some non-negligible volume, means that his model and the ideal gas model cannot be applied simultaneously by one person to one and the same state of affairs. In present terms this is to say that they are perspectivally incompatible and cannot be treated logically simultaneously in application to the same intended domain. And this is so even if in their respective applications they were to suggest the same results.[9] But this is not to say that the two models contradict one another, or that they are in any sense incompatible when viewed as unapplied systems. First, they do not contradict each other since neither is a linguistic entity, let alone a proposition – i.e. neither model asserts anything to be the case. And second, while formal relations might be established between them, the most that such formal relations could be hoped to show is that they are different, not that they conflict. Viewed as unapplied systems, the models, in a sense, stand side by

---

[8] The notion of reduction suggested here is essentially similar to the notion of *correspondence* treated e.g. in Krajewski (1977), Ch. 1: cf. esp. pp. 6 and 10. Note that the present notion is not intended to be the one employed when speaking in such contexts as that concerning e.g. the reduction of biology to physics.

It might also be noted at this point that the ideal gas model and that of van der Waals, in being conceptually distinct, strictly speaking constitute the respective bases of independent theories, though both fall under the more general heading of the kinetic theory of gases.

[9] Imagine, for example, Bernoulli's model and the ideal gas model to give identical results: they would nevertheless be perspectivally incompatible due to the assumption of an infinite number of molecules in the former and a finite number in the latter.

side, and neither impinges on the domain of the other. It is only in their application that conflict arises.

In the simpler case treated in the previous chapter, it was suggested that two predicates would conflict in application to one and the same thing only if both predicates were of the same category, or, in terms of the present chapter, only if they were different values of the same parameter. And in the present case we should say that two models would conflict in application to the same intended domain only if both models involve the same parameters. Furthermore, as in the previous chapter, evidence concerning whether or not different models do involve the same parameters may be afforded via a consideration of whether their actual application or testing requires the same (sorts of) operations or measurements. Thus the fact that the application of both the van der Waals model and the ideal gas model involves identical operations using a ruler (or a more sophisticated instrument), a manometer, and a thermometer suggests that both models involve the parameters volume, pressure and temperature.

In the present case not only does the testing or application of the models under consideration require the same sorts of operations involving the same sorts of instruments, but in certain applications the models suggest measurably different results of their common operations. When applied to gases at high pressures and low temperatures van der Waals' model, or its equation of state employed as the expression of a law, predicts values much closer to those obtained on measuring instruments than does the ideal gas model (Boyle's law, or the general gas law). van der Waals' model is thus more *accurate* than the ideal gas model, and, other things being equal, we should consequently say that it is the more *acceptable* of the two, and thus that it constitutes a scientific *progression* beyond the other. This conclusion is further supported by the fact that the actual *scope* of van der Waals' model is much wider than that of the ideal gas model, for it suggests results that are at least of the correct order of magnitude in the case of change of state, while this case lies quite beyond what is tractable on the ideal gas model. On the other hand, the ideal gas model is *simpler* than that of van der Waals; this fact, plus the relatively adequate results obtained by Boyle's law in the case of gases at low pressures and high temperatures, prevents its being discarded from science.

6. THEORIES AS DISTINCT FROM LAWS

One of the major shortcomings of modern philosophy of science as a whole, and of the logical empiricist and Popperian views in particular, has been its failure to provide a positive characterisation of the essential difference between the natures of scientific laws and theories. On the present view however, as has been mentioned above, scientific theories are conceived to be applied models; and the expression of a (quantitative) scientific law, rather than take the form of a statement as on the empiricist and Popperian views, is to consist of an *equation* relating certain *measurable parameters*. This conception of laws, and with it the role here seen to be played by parameters and measuring instruments, is in keeping with Reinhold Fürth's characterisation, where he says:

It is the generally accepted view [of physicists] that the laws of physics are expressed in the form of mathematical equations between certain variable quantities or 'parameters' which may either be capable of assuming any values within a certain range (continuous parameters) or are restricted to a finite or infinite, but enumerable set of discrete values (discontinuous parameters). A parameter can either be directly defined 'operationally,' that is by a well defined process of measurement on a given physical system by means of certain measuring instruments, the readings on which determine the numerical values of the parameter; or it may be defined indirectly by a mathematical formula in terms of other directly defined parameters.[10]

On the present view, following the above example, scientific laws are seen to be of essentially two sorts, where one and the same law may qualify as being of both sorts. Thus, for example, Boyle's law may be arrived at solely on the basis of the performing of certain operations with measuring instruments, or, it may be arrived at via derivation from the ideal gas model. In the first case it is considered an *empirical* or *experimental* law,[11] and, in the second, it will here be

---

[10] Fürth (1969), p. 327. In the present regard, cf. also Poincaré (1902), p. 217, and Campbell (1920), Ch. II.

[11] Cf. ibid., p. 153: "[W]hy do we call some laws 'empirical' and associate with that term a slight element of distrust? Because such laws are not explained by any theory." An empirical law is here taken as not necessarily involving measurements – we might call such a law as does a *quantitative* law; and an experimental law is simply to be one which applies in experimental situations.

termed a *theoretical* law.[12] Furthermore, it may be the case that certain laws are only empirical, in that they have as yet not been successfully derived from any model or theory; or they may be only theoretical, in that they have been derived from some model, but, perhaps due to the absence of sufficiently sensitive measuring instruments, have not been confirmed by experiment. In this way then we should say that van der Waals' law is more a theoretical law than an empirical one, for, while it has received a certain amount of support from empirical investigations, it takes the particular form that it does as a result of the nature of the theory or model from which it has been derived, and empirical investigations alone would not have suggested its having exactly that form.

Scientific theories, on the other hand, while they may in certain cases be *expressed* by equations (relating the constituents of a model), are here positively characterised as being applied models – i e models from which empirical laws can be derived. Thus the present view is very much in keeping with that of N. R. Campbell where, in summary, he says:

[T]he value of the [dynamic theory of gases] is derived largely, not from the formal constitution, but from an analogy displayed by the hypothesis.

-----

[12] This usage of the term differs from that of e.g. Carnap and Nagel, in which a theoretical law is necessarily to contain terms referring to unobservables. See e.g. Nagel (1961), p. 80, and Carnap (1966a), p. 227. A discussion of the issues raised in the present chapter may also be found in Hempel (1970). In these later writings, each of the above authors makes reference to Campbell, and their respective discussions are largely shaped by Campbell's (1920). Hempel in fact goes so far as to suggest that Campbell, who emphasises so strongly the notion of analogy, is a proponent of what Hempel calls the 'standard conception' of scientific theories. The fact is that Campbell's view lies quite beyond the standard conception, if by this we understand Hempel to mean the logical empiricist conception, or anything closely resembling it. (Consider, e.g., Campbell's saying: "Of course the province and power of logic have been very greatly extended in recent years, but some of its essential features ... have remained unchanged; and any process of thought which does not show those features is still illogical. But illogical is not synonymous with erroneous. I believe that all important scientific thought is illogical, and that we shall be led into nothing but error if we try to force scientific reasoning into the forms prescribed by logical canons." 1920, p. 52.) What the above authors are actually doing in the works cited here is not so much providing elaborations of the logical empiricist conception of theories as affording relatively neutral descriptions of the way theories function in science, descriptions which must be seen as being ad hoc in relation to their philosophies of science.

This analogy is essential to and inseparable from the theory and is not merely an aid to its formulation. Herein lies the difference between a law and a theory, a difference which is of the first importance.[13]

Now, in considering theories to be applied models, the present study is by no means suggesting that such models must be either picturable or mechanical; and part of what is to follow is based on the fact that they seldom, if ever, are the former.

Pierre Duhem, in the chapter of his book *The Aim and Structure of Physical Theory* entitled 'Abstract Theories and Mechanical Models,' has argued against the necessity of employing mechanical models in scientific theorising. To the extent that his remarks are taken to suggest that *no* models are required in theorising, the present study must of course object. But a closer reading of his text suggests that what he is actually arguing against is the need of employing *picturable* models; and if this is so, there is no disagreement between his view and the present one.

Duhem begins his argument by distinguishing between the 'abstract mind' of the Continental physicist and the 'visualising mind' of the English physicist; and he suggests that men possessing minds of the former sort "have no difficulty in conceiving of an idea which abstraction has stripped of everything that would stimulate the sensual memory,"[14] while the intellectual power of the latter "is subject to one condition; namely, the objects to which it is directed must be those falling within the purview of the senses, they must be tangible or visible." And at a later point he says, "The French and German physicist conceives, in the space separating two conductors, abstract lines of force having no thickness or real existence; the English physicist materializes these lines and thickens them to the dimensions of a tube which he will fill with vulcanized rubber."[15]

Now even if it is assumed that the ultimate aim of scientific theorising is to produce a mechanical model of the sort of phenomenon under investigation, such a model need not be one of the sort which apparently concerns Duhem – i.e. it need not be a picturable one. Thus, for example, we might have a model employing abstractions such as the idea of a mass-point – which cannot be pictured – while

---

[13] Ibid., p. 119; cf. also pp. 129–132.
[14] This quote and the next are from Duhem (1906), p. 56.
[15] Ibid., p. 70.

the model is nevertheless mechanical. (The same is true of the ideal gas model if we take its molecules as having no volume, rather than merely negligible volume.) But what is of importance here is that on the present conception scientific theories, while each is essentially related to some sort of model, need not be related to visualisable models, and as a matter of fact almost always involve abstractions or *idealisations* (such as that of a mass-point) in their attempts to provide an understanding of the essence of certain natural phenomena.

7. IDEALISATION

In recent years a number of Polish philosophers have drawn attention to the importance of idealisation in science, and to the fact that its existence runs counter to the conception of science suggested by the more traditional views. In this regard Leszek Nowak, for example, has said:

Let us notice that the basic assumption of the inductivist model of science is the idea of science as a form of registration, or phenomena systematisation. ...

This idea is not beyond controversy, however. Let us take a look at the theoretical activities of a physicist. His basic activity is the construction of models of different sorts: the model of the perfectly rigid body, the model of the homogeneous cosmos, the model of the perfect gas, etc. The construction of such models always involves the omission of certain features of material objects ... . Thus, where real physical objects have three spatial dimensions, plus mass, acceleration, etc., their model, the material point, has zero dimensions, and only certain (mass, acceleration, etc.) characteristics of the objects so modelled .... Model construction is, then, anything but phenomena registration – the model is not "an abbreviated record" of a phenomenon but its "deformation."[16]

Nowak goes on to suggest that the viability of the model rests on whether the characteristics emphasised in it are *essential* to the phenomenon.[17]

In his book *Correspondence Principle and Growth of Science*, Władysław Krajewski has provided many examples involving idealisation (including the ideal gas model and that of van der Waals)

---

[16] Nowak (1979), pp. 284–285.
[17] In this regard see also Hesse (1966), pp. 34f.

taken from both the physical and social sciences; and he points out how these sorts of cases differ from the 'crude empiricism' of Aristotle, in which idealisation does not play a role.[18] Both Nowak and Krajewski have recognised that the employment of idealisation involves the making of *counterfactual* assumptions, and that it consequently creates serious problems for the empiricist conception of scientific laws and theories. Nowak has also suggested that idealisation is not in keeping with Popper's philosophy of science; and in the present study the reason for this becomes clear, namely, that the Popperian conception, similarly to the empiricist, is based on the Deductive Model (which, incidentally, closely resembles the Aristotelian syllogism, as has been noted in Chapter 1).

The present conception, on the other hand, recognises the importance of idealisation in science, and sees it as playing a central role in the determination of the natures of individual scientific theories. And, by taking scientific theories to be applied models, it is an easy step to the view that it is precisely such ideal entities as mass-points and perfectly elastic spheres, occurring in these models, that are the referents of theoretical terms.

8. THEORETICAL TERMS AND CORRESPONDENCE RULES

In Chapter 4 the notions of theoretical terms and correspondence rules were discussed in the context of the empiricist conception of the relation between theories and laws, and were found to create grave problems for the empiricist view.

Theoretical Terms

The logical empiricist problem of theoretical terms concerns the *meaningfulness* of such terms, and results from the fact that on the empiricist conception the meanings of all non-logical terms in science are ultimately to be determined on the basis of their formal relation to observational terms. And, in keeping with its logical positivist beginnings, this view sees observational terms as being those the

---

[18] Cf. Krajewski (1977), Ch. 2. For the present author's view on idealisation, as expressed elsewhere, see Appendix IV below and, for greater detail, Dilworth (2007), pp. 123–127.

referents of which are entities of the sense-data variety. As suggested in Chapter 4, this requirement creates problems not only with regard to terms of the sort mentioned in the previous section, but also with regard to mensural terms, and terms the referents of which have dispositional properties.[19] Unfortunately, due to the similar intractability of terms of this latter sort on the empiricist conception, they have also come to be called theoretical, thus confusing the distinction between empirical or experimental laws on the one hand, and scientific theories on the other.

On the present conception, however, science is not seen to rest ultimately on that which can be directly experienced – subjective sense-data, but on the results of operations involving the employment of certain instruments – results which can be intersubjectively shared.[20] Thus mensural notions such as that of temperature are not viewed as being theoretical (which is not to say that they are free of a priori presuppositions), but as being empirical or experimental, since they are employed in the expression of empirical laws which, taken together, may be seen to constitute the whole of the empirical aspect of science.

The present view also differs in important ways from that of P. W. Bridgman, though he too emphasises the role played by operations in science. The major difference between his view and the present one lies in the stress he places on the *meanings* of scientific terms, and with it his suggestion that their meanings are ultimately determined by the operations performed in their application.[21] Here, on the other hand, meanings do not play a central role, but rather the operations do so themselves. Thus, as has been suggested earlier, two different theories employing terms with quite different meanings may both be related to the same operations. But of greater importance at this point is that on the present view theoretical notions are not seen to depend on operations in order to be meaningful, and it is in fact often the case that an antecedent appreciation of their 'meanings' suggests which operations may be performed to determine their relative applicability.

---

[19] For a discussion of dispositional properties which is in keeping with the view of the present study, see Agazzi (1976), pp. 149ff.

[20] With regard to the relation between the performing of operations and intersubjectivity in science, see Agazzi (1977a), pp. 162ff., and (1978), pp. 100ff.

[21] Cf. e.g. Bridgman (1936), Chh. II and IV.

Thus the present view makes a clear departure from both Bridgman and the empiricists by suggesting that theoretical terms seldom, if ever, obtain their meanings from the observational basis to which the theory is intended to apply,[22] but obtain them rather from any source whatever – sources as varied as metaphysics and everyday discourse.

Thus, in the case of the kinetic theory of gases, we see that the basic theoretical notions are obtained from the metaphysics of atomism (though they would have been just as 'meaningful' if newly introduced), and from such everyday concepts as random motion and elasticity. And while the molecules in an ideal gas are to move in a way which is in keeping with Newton's laws of motion, the meanings of the terms used in describing the model can hardly as a consequence be said to be dependent on those laws, much less on the fact that the laws might themselves be considered empirical.

Correspondence Rules

As mentioned in Chapter 4, the logical empiricist problem of how theoretical terms obtain their meanings is closely linked to their problem of how empirical laws can be formally derived from theories containing such terms. And this latter problem also arises for the Popperian view, for, in being based on the Deductive Model, it too requires a formal derivation from the theoretical level to the empirical. In recognising the impossibility of a direct derivation, empiricists have noted that in actual science certain 'rules' (correspondence rules) are required in order to connect theoretical notions to empirical ones. (The question of whether, given these rules, the resulting derivation is a strictly formal one will here be left open.) Thus, for example, Hempel has said:

In the classical kinetic theory of gases, the internal principles are assumptions about the gas molecules; they concern their size, their mass, their large number; and they include also various laws, partly taken over from classical mechanics, partly statistical in nature, pertaining to the motions and collisions of the molecules, and to the resulting changes in their

---

[22] For a lucid critique which is of relevance to this and other points raised in the present chapter, see Spector (1965). Concerning the present point, see also Kuhn (1974), pp. 465–466.

momenta and energies. The bridge principles [correspondence rules] include statements such as that the temperature of a gas is proportional to the mean kinetic energy of its molecules, and that the rates at which different gases diffuse through the walls of a container are proportional to the numbers of molecules of the gases in question and to their average speeds.[23]

On the present view however – where models are taken to play a central role in theorising – rather than posing a problem, correspondence rules are seen to be precisely what is required in order for the analogy displayed by the model to be applicable to empirical situations. In contrast with this, the admission of their existence on the part of empiricists is at the same time an admission that empirical laws cannot be derived directly from abstract theories – i.e. that the derivation is indirect, since it depends on these rules. That the derivation is indirect is made all the clearer when we realise that the correspondence rules are often only tacitly given, and that their nature is not dictated by the model (theory) itself. As regards this latter point, we see that such rules are more like conventions determined by scientists in order that the theory *be* applicable to certain empirical states of affairs. Thus we also see that one and the same model, when applied to different sorts of situations, will have different correspondence rules. On the present view we should thus say that the correspondence rules employed in the case of applying a certain model are determined partly by the nature of the model itself, and partly by the *intention* of the person(s) applying it. And to this we may add that the application of the model is to consist in the derivation of its empirically testable theoretical laws with the aid of just such rules.

9. REALISM VS. INSTRUMENTALISM

On the present view the debate between realism and instrumentalism with regard to the ontological status of theoretical entities is left open. It may be pointed out however that the present view suggests that to the extent that theoretical notions are idealisations or abstractions, their real counterparts, should they exist, cannot have exactly those properties supposed by the notions. For example, granting that real gases are composed of molecules in rapid motion, these

---

[23] Hempel (1970), p. 144. See also e.g. Nagel (1961), pp. 93–94.

molecules could not have zero volume (as one interpretation of the ideal gas model might suggest), nor, it seems, could they be perfectly spherical or perfectly elastic.[24]

On the other hand, to deny that they exist at all might also appear problematic given that we understand theories as providing explanations either of empirical laws or of individual phenomena. A natural interpretation of the role of theories and laws suggests that while an empirical law taken by itself only evinces a form of constant conjunction, in certain cases when derived from a theory it might attain the status of a causal law. For example, having derived Boyle's law from the ideal gas model we might say that the pressure of a gas increases with decreasing volume because the number of molecules striking the walls of its container is thereby increased per unit time, and that in this way we have consequently explained the phenomenon. We have explained that which at first we did not understand: the constant conjunction, in terms of that which we do: the model.[25] And it might be thought that to accept this view requires some sort of commitment to the existence of molecules in real gases.

Now, while the above conclusion might itself be challenged, what is being stressed here is simply that the present view excludes neither a realist nor an instrumentalist approach to the nature of scientific theories.

The development of the present conception given in this chapter should not only have helped clarify its connection to actual science, but it should also have revealed further difficulties faced by similar

---

[24] In this regard cf. Boltzmann (1896), p. 26, where he says: "In describing the theory of gases as a mechanical analogy, we have already indicated, by the choice of this word, how far removed we are from that viewpoint which would see in visible matter the true properties of the smallest particles of the body."

[25] It might thus appear that we would have an instance of what Campbell calls 'explanation by greater familiarity,' where he suggests that "The theory of gases explains Boyle's Law, not only because it shows that it can be regarded as a consequence of the same general principle as Gay-Lussac's Law, but also because it associates both laws with the more familiar idea of the motion of elastic particles." (1920), p. 146. We should suggest however that it is not the greater familiarity of the idea of the motion of elastic particles that allows the theory of gases to explain the gas laws, but the *understanding* afforded by the fact that the idea involves material objects between which there are to exist causal relations.

With regard to the idea of models characterising the essence of phenomena, and thereby causally explaining them, see also Nowak (1980), pp. 128–129.

attempts to apply the empiricist and Popperian conceptions. Of major importance in this regard is the latters' consideration of theories as being universal or general statements which are formally to entail the descriptions of certain sorts of phenomena, whereas in actual fact, as is suggested by the present view, theories are more like models which bear, not a logical relation to (the descriptions of) phenomena, but an analogical relation to them. In the next chapter an alternative account to the present one will be considered, in which a notion of model is also to play a fundamental role.

CHAPTER 11

THE SET-THEORETIC CONCEPTION OF SCIENCE

1. A NEW FORMAL APPROACH TO SCIENCE

In recent years a new view has emerged in the philosophy of science, taking as its basis the informal axiomatisation of Newtonian particle mechanics in terms of a set-theoretical predicate. This axiomatisation itself appears first in McKinsey, Sugar, and Suppes (1953), is employed in Adams (1959) in such a way as involves particular notions of reduction and intended model, and, in an attempt to handle theory dynamics, has been further developed by J. D. Sneed in his book *The Logical Structure of Mathematical Physics* (1971). In this book Sneed attempts to reconstruct Newtonian particle mechanics in such a way as to clarify the role of theoretical terms in science, and to provide a conception of how scientific theories can rationally evolve in the face of recalcitrant data.

The above approach, constituting the set-theoretic or 'structuralist' conception, is similar to the Perspectivist conception to the extent that it does not see scientific theories as being universal statements. On the other hand however, it is in a more fundamental respect similar to the empiricist and Popperian views, for its basis is a particular *formal* system, namely, intuitive set theory.

In the present chapter a critique will be made of the set-theoretic view as it is presented by Sneed, and by Wolfgang Stegmüller.[1] A key feature distinguishing Sneed's approach from that of his predecessors is the introduction and extensive treatment of what Sneed calls 'the problem of theoretical terms.'

2. SNEED'S PROBLEM OF THEORETICAL TERMS

As discussed in Chapters 4 and 10 above, the empiricist problem of theoretical terms has its origin in the logical positivist criterion of

---

[1] Cf. Stegmüller (1973).

empirical meaningfulness: that a non-formal statement is meaningful only if it is verifiable by direct observation. And, as has been pointed out in those places, this problem arises not only in the case of terms such as "mass-point" and "electron," but also in the case of such mensural terms as "mass" and "temperature," as well as in the case of terms such as "magnet," whose referents have dispositional properties; and it has led to the classification of these latter sorts of term as also being theoretical. On the empiricist conception of science then, if a solution to this problem of 'theoretical' terms were to be found, it would have to lie in the direction of reducing such terms to terms the referents of which are directly observable.

Sneed's problem of theoretical terms, on the other hand, is quite different, for it does not concern the empiricist conception, but the set-theoretic one; and, of the three sorts of term mentioned above, it pertains only to those of the mensural variety.

Sneed's Notion of Theoretical Term

For Sneed, a theoretical term is a function any measurement of the value of which implies that the theory in which the function occurs has already been successfully applied.[2] Thus the question as to whether a particular function is theoretical must be relativised to a particular theory. Where a function is theoretical relative to a theory θ, it is called θ-dependent. Sneed takes the mass and force functions of Newtonian particle mechanics to be functions of this kind, while, for example, the position function is not. In evaluating his concept of theoretical term we might thus consider the example he gives as his main support for treating "mass" as theoretical relative to classical particle mechanics:

> An example of a θ-dependent function is the mass function in an application of classical particle mechanics to a projectile problem. In this case we typically determine the mass of the projectile by "comparing" it to some standard body with a device like an analytical balance or an Atwood's machine. ... But the only reason we believe that these comparison procedures yield mass-ratios, and not just numbers, completely unrelated to classical particle mechanics, is that we believe classical particle mechanics applies (at least approximately) to the physical systems used to make the

---

[2] Cf. Sneed (1971), pp. 31ff. and 116f.

comparisons. If someone asks why the number $(a/g - 1/a/g + 1)$, calculated from the acceleration observed in an Atwood's machine experiment, is the mass-ratio of the two bodies involved, we reply by deriving it from the application of classical particle mechanics to this system. I maintain that examination of any acceptable account of how the mass of a projectile might be determined would reveal the same sort of dependence on an assumption that classical particle mechanics applied to the physical system used in making the mass determination.[3]

To begin, we might ask to what extent this example supports Sneed's claim that "mass" is theoretical in his sense. While it may here be granted that the use of an Atwood machine to determine relative masses or 'mass-ratios' does presuppose the viability of Newton's second law, we note that what Sneed has described is not actually an experiment (test), nor is it a determination of the sort the machine is usually employed to make (which is that involving the establishment of g); what is more, both this latter use and that to which the machine was originally put (the results of which should allow it to be employed in the way described by Sneed) presuppose that the mass-ratios in question are already known. Thus, with regard to its original use for example, we read in Hanson: "Atwood showed that when $m_1 = 48$ gm. and $m_2 = 50$ gm. then the acceleration of m_2 is indeed 20 cm./sec.^{2}. The results were carefully recorded and generalised: they squared with the predictions of the second law. For Atwood this fully confirmed the law."[4] Taking Hanson's description to be correct, how then did Atwood determine the values for m_1 and m_2? He could not have used his machine to do so, on pain of circularity. He most probably measured these masses directly, using a beam balance, or some similar device. Thus, to say the least, Sneed has chosen a rather peculiar example to support his claim that *every* measurement of mass presupposes that classical particle mechanics applies to the system used in making the mass determination. He also makes reference however to an 'analytical balance,' which is to be some kind of pan balance, and maintains that such a balance, like Atwood's

---

[3] Ibid., pp. 32–33. (We take this occasion to remark that an Atwood machine is essentially a pulley over which is passed a thread whose ends are connected to weights of unequal mass.)

[4] Hanson (1958), p. 101.

machine, constitutes a set-theoretical model of Newtonian mechanics.[5] But, apart from the fact that Atwood's machine and 'analytical balances' do *not* constitute such models – for reasons to be given below – this is not to say that the use of such a balance *presupposes* Newtonian mechanics. Of course it does not presuppose Newtonian mechanics, since this sort of balance was in use long before the time of Newton. Thus, on the basis of his examples, Sneed is hardly justified in claiming "mass" to be theoretical in his sense.

Noting this then, and that the same sort of comment applies to Sneed's treatment of force – since he does little more than simply claim its measurement to presuppose particle mechanics, we might go on to consider whether, even if he had shown "mass" and "force" to be theoretical in his sense, his sense is one which can be generally accepted as capturing the essence of what theoretical terms actually are. In this regard we note first that Sneed does not seem to see scientific laws as being anything other than the constituents of scientific theories, and more often than not in discussing the referents of what he considers to be theoretical terms he speaks of their measurement presupposing some law. Thus, for example, in giving his reason for considering "force" to be theoretical, he says: "All means of measuring forces, known to me, appear to rest, in a quite straightforward way, on the assumption that Newton's second law is true in some physical system, and indeed also on the assumption that some particular force law holds."[6] However, while it is not being claimed in the present study that Newton's second law is a straightforwardly empirical one, all the same it is one thing to say that measurements of a certain sort depend on a particular *theory* (classical particle mechanics), and quite another to say that they rest on a particular *law*, even if the law is intimately related to the theory. And it seems strange to call e.g. "force" theoretical because its measurement should presuppose some law. This point concerns more than mere terminology, for there are views of science in which measurements

---

[5] Sneed (1971), p. 117.
[6] Ibid.

are based on laws, not on theories, and in which, for this reason, no mensural term is to be considered theoretical.[7]

More generally, when we consider the extent to which Sneed's distinction between what he calls theoretical and non-theoretical terms serves to clarify this distinction as it occurs in actual science, we see that not only are Sneed's theoretical terms mensural terms, but his distinction provides no positive understanding of what an empirical term should be. And to this it must be added that his notion of theoretical term does not pertain at all to terms of the sort as are normally considered to be particularly exemplary of theoretical terms, e.g. terms such as "mass-point" and "electron," the referents of which are unobservable.

The present critique of Sneed's notion of theoretical term is not intended to suggest, however, that the notions of mass and force in Newtonian mechanics are unproblematic, for there of course exist the age-old problems of providing an intuitively satisfying definition of "mass," and of deciding whether Newton's second law ought to be considered as anything more than merely a definition of "force." Nor is it the present intention to suggest that in testing a particular theory, or a law, there do not occur situations in which the theory or law being tested must in some sense, or to some extent, itself be presupposed in order to carry out the test.[8] But the very existence of these sorts of case draws attention to a deeper issue regarding the viability of Sneed's notion of a theoretical term, namely, that he introduces and almost always employs the notion in the context of his own conception of science, and it is this use of it which he ultimately must justify. Thus, for example, even if he were to succeed in showing that in actual science any measurement of mass in some sense presupposes the applicability of classical particle mechanics, he would still have to show that the essence of this phenomenon is retained when we move over to thinking of 'mass' as a function occurring in the definition (or extension) of a set-theoretical predicate, and the

---

[7] Cf. e.g. Campbell (1928), p. 1: "[T]rue measurement, as we shall see ... is based upon ... scientific laws. We shall be nearer the truth if we define measurement as the assignment of numerals to represent properties *in accordance with scientific laws.*"

[8] For a discussion of this problem (in Swedish) with regard to Ohm's law, see Johansson (1978); concerning the conventional aspects of Newton's laws of motion, see Ellis (1965).

successful application of a theory as consisting in the discovery of a set-theoretical model for such a predicate.

Sneed's Problem of Theoretical Terms

Sneed presents his 'problem of theoretical terms' as a problem which can arise, given his notion of theoretical term, for a relatively straightforward construal of his own view of science. On Sneed's view, i.e. on the set-theoretic conception, a *theory* (of mathematical physics) is to consist of a 'formal mathematical structure' characterised by a set-theoretical predicate, plus a set of physical systems each of which is an 'intended application' of the theory.[9] An *empirical* claim of the theory is a statement to the effect that the predicate is in fact applicable to one of its intended applications, i.e., that one of these physical systems is a set-theoretical model of the predicate. And for Sneed the problem of theoretical terms is one of providing a conception of how the truth-value of such a claim may be determined given that the definition of the predicate might well contain terms which are theoretical in his own sense.

Sneed takes the possible existence of this problem for the set-theoretic view as being sufficient motivation for the rest of his discussion, and, presupposing this view, devotes a major part of his book to reasonings in terms of his theoretical/non-theoretical distinction. His efforts in attempting to solve his problem involve modifications on the method (discussed in Chapter 4) taken over by the empiricists from Ramsey in an attempt to solve their problem of theoretical terms. Unfortunately, however, in spite of his lengthy development of 'Ramsey eliminability,' the results he obtains seem to apply in a

---

[9] Cf. Sneed (1971), pp. 36 and 161. In its simplest form a theory is to be, more particularly, an ordered pair, the first element of which is the set of models of the predicate, and the second element of which is the set of those empirical states of affairs which it is intended should be models of the predicate: cf. Adams (1959), p. 259. Though in his book Sneed makes of the first element an ordered *n*-tuple containing sets of possible models, possible partial models, constraints and so on, the essence of the approach outlined in Adams (1959) is retained. Both elements in the set-theoretic notion of a theory run counter to one's intuitions: the first because on Sneed's reasoning any model − physical or otherwise − of the predicate should constitute a part of the theory's formal mathematical structure, and the second because those states of affairs to which it is intended that the theory be applied are conceived as being part of the theory itself.

straightforward way only when the values of the 'theoretical' functions occurring in the 'formal mathematical structure' are in principle determinable by computation from the values of the 'non-theoretical' functions; and, in the case of particle mechanics, Sneed in the end concludes that "it is not likely that the mass function can be Ramsey eliminated from *any* claim of classical particle mechanics."[10]

In any case, whether Sneed has or has not overcome his problem of theoretical terms is of little consequence to the present critique, for his problem is couched in the context of his own conception of science, and it is this conception which is here being called into question. But, from the present point of view, what is of importance here is the fact that both Sneed's theoretical/non-theoretical distinction as employed in the context of the set-theoretic conception, and his problem of theoretical terms, assume as a fundamental presupposition that the models of set-theoretical predicates characterising theories of mathematical physics may be found in the real world — that it is through the determination of just such models that these theories are to apply to empirical reality. As the present chapter proceeds, however, the attempt will be made to show that this idea is quite mistaken.

As mentioned above, on the set-theoretic view a scientific theory is to consist of a certain 'mathematical structure' and a set of 'intended applications,' the latter being the set of real states of affairs to which it is intended that the theory apply. But, as will now be shown, one indication of the incorrectness of thinking of the models of predicates characterising scientific theories as existing in the real world lies in the inability of the set-theoretic conception to delineate this set of 'intended applications' so long as it restricts itself to employing the methods afforded by set theory.

3. THE PROBLEM OF DELINEATING THE INTENDED DOMAIN

In attempting to identify the realm of intended applications of a theory on the set-theoretic view, we might begin by considering as a candidate the set of (set-theoretical) models of the predicate used to characterise the mathematical structure of the theory. Were we to do

---

[10] Sneed (1971), p. 152.

this however, we would quickly see that the adoption of this set of models, i.e. this set of sets to which the predicate correctly applies – the predicate's *extension*, would leave us saying that the theory is intended to apply to all and only those states of affairs to which it in fact does apply; and this of course would not at all capture the notion we are after.[11] In an attempt to overcome this difficulty, Sneed conceives of those states of affairs to which it is intended that a theory be applied as being a subset of the set of models to which a sub-part of the predicate applies. Of course the theory's extension is itself such a subset, and so the theory's set of 'intended applications' requires further distinguishing characterisation.

At this point Sneed (as well as Stegmüller, though his treatment of the case is slightly different)[12] quite transcends the methods afforded by set theory in filling out his picture of science. Using the notation of the set-theoretic conception, we have at present that the set I of intended applications of a theory is a subset of the set M_{pp} of models of a particular part of the predicate characterising the theory's mathematical structure (this part of the predicate is not to contain definitions of 'theoretical' functions, and its models are thus to be 'empirically' recognisable as such); i.e., we have that $I \supseteq M_{pp}$, where M_{pp} is the set of models of the restricted predicate. We note that we also have $M \supseteq M_{pp}$, where we here take M to be the set of models of the complete predicate. And the question is then: how is I to be characterised so as to distinguish it from M?[13]

Sneed here turns to the *beliefs* of individual persons, e.g. physicists, and suggests that there exists for them a set of *paradigm examples* (à la Wittgenstein) $I_0 \supseteq I$ each member of which is believed to be extendible so as to constitute an element of M, and which furthermore characterises I in that all members of I bear a relation of the

---

[11] For a discussion of this point and others relevant to the present critique, see Hempel (1970), pp. 150ff.; with regard to the problem of delineating the intended domain see also Agazzi (1976), pp. 144–145, and Nagel (1961), pp. 93ff.

[12] Cf. Stegmüller (1973), pp. 193ff.

[13] With regard to this question we note that there exist set-theoretical models of e.g. classical particle mechanics to which it would normally never be intended that the theory be applied. Cf. e.g. Adams (1959), pp. 258–259, where the set P of 'particles' in a model of the theory has the number 1 as its sole member.

same sort as Wittgenstein's 'family resemblance' to the members of I_0.[14]
I is further characterised by requiring that it meet certain other conditions, one of which is that its members be physical systems (though
Sneed is unable to tell us what is to count as a physical system on his
view);[15] and instances of paradigm examples – members of I_0 – in the
case of classical particle mechanics are suggested to be the solar
system and its various subsystems.[16]

The point to be made here is similar to those raised in Chapter 6
with regard to Popper's and Lakatos' transcendence of the Deductive
Model. Irrespective of the extent to which the above description does
or does not capture the way in which the intended domain of application of a theory is determined by physicists, by referring e.g. to their
(potentially vacillating) beliefs it reaches beyond what is warranted
on the basis of a conception of scientific theories in which they are
formulated in terms of set-theoretical predicates, and in this way is
quite ad hoc.[17] Not only this, but such a notion as that of characterising a set by reference to some vague trait similar to that of 'family
resemblance' in fact runs counter to the spirit of intuitive set theory,
on which Sneed ultimately bases his reasoning.

Thus we see that the set-theoretic approach has been unable to
provide a coherent conception capable of accounting for the way in
which scientific theories are related to those states of affairs to which
they are *intended* to apply. And, as will be seen below, the concepts
afforded by set theory function no better when it comes to handling
the nature of the relation between scientific theories and those real
states of affairs to which they are thought to *succeed* in applying.

4. THE PROBLEM OF EXTENSION

In its simplest form, the problem faced by the set-theoretic conception with regard to extension is that there *are* no physical models

---

[14] Cf. Sneed (1971), pp. 268–270.

[15] Ibid., p. 250.

[16] Stegmüller (1973), p. 175.

[17] A similar charge can also be levelled both at Sneed's definition of 'having a theory'
(1971, p. 266), which presupposes such indistinct notions as 'a person having observational evidence,' and at Stegmüller's development of Sneed's views in 'pragmatic' or
'extra-logical' terms: cf. e.g. Stegmüller (1976), p. 154, and (1973), p. 162.

which satisfy the mathematical structure of a scientific theory axiomatised by a set theoretical predicate. For example, the solar system is not a model of Newtonian particle mechanics, evidence for which fact is, e.g., the inability of Newtonian theory to account for the rate of the advance of the perihelion of Mercury. And, more generally, even in cases where theories are thought to apply quite well, they do not apply exactly, as they must in order for that to which they are being applied to constitute their model in the set-theoretical sense.[18]

But an even stronger claim can be made here: it is that no physical system *could* be a set-theoretical model of any modern scientific theory. And the reason for this is that such theories, by their very nature, involve *idealisations* (as discussed in the previous chapter) the referents of which could not exist in empirical reality. Sneed himself realises the existence of this problem for his view:

It might be argued that one simply cannot expect to find 'real' physical models for S, or anything like it. If we limit ourselves to domains consisting of physical objects or other non-abstract individuals (e.g. sounds or events) which we can, in some way, perceive, and define functions in terms of "observable" relations among these objects, then ['Q is an S'] will always (or almost always) be false. Speaking platonistically, S is an ideal of which real objects are, at best, imperfect copies.[19]

Sneed then sketches two ways this problem might be approached – each of which would require a treatment transcending what could be provided by the concepts of orthodox set theory – and admits that he does not see how either way is to be worked out in detail. But he then suggests that the questions of logical structure he intends to examine – namely, those concerning the use of set-theoretical predicates characterising scientific theories to make empirical claims – can be considered independently of this matter. But can they? If the

---

[18] The attempt has been made in Moulines (1976) to overcome this sort of problem via the introduction of the notion of a 'fuzzy set.' But even aside from the fact that this move is not at all indicated by the set-theoretical basis Moulines assumes, it involves the counterintuitive assumption that scientific theories themselves are inexact in their predictions, rather than that their predictions, while exact, are not identical with the results obtained in the performance of actual measurements.

[19] Sneed (1971), p. 24; cf. also Sneed's remarks on pp. 112f. and 120. Note that the point being raised here concerns those sorts of entities which *could* not exist in the real world, not those which, while they might exist, would be impossible to detect.

set-theoretical predicates representing scientific theories can in principle only have abstract models, then no state of affairs of which
Sneed's 'empirical claims' may be true could be an empirical one.
And all of Sneed's reasoning concerning the role of what he calls
theoretical terms in the making of what he calls empirical claims
would have no bearing on the question of the nature of the relation
between scientific theories and empirical states of affairs. It is here
suggested that this is in fact the case: that the models of modern scientific theories characterised by set-theoretical predicates are such
that they could not exist in the empirical world, and that as a consequence the major portion of Sneed's work bears little relation to actual science, since it is based on the belief that empirical states of
affairs might, at least in principle, constitute such models.

The fact that the set-theoretical models of predicates used to characterise scientific theories themselves involve idealisations suggests
rather that such 'models' may be thought to be of the same sort as
the (ideal) models treated in the previous chapter. In this case one
might thus apply oneself to characterising such ideal models in science
in terms of set-theoretical predicates. But it may be noted that such
an exercise in formal reconstruction would at most succeed in presenting these models in a new form, and could in itself be expected
to contribute little towards our understanding of the role that they
play in science.

The considerations of the last two sections reveal then that the
employment of the tools of orthodox set theory is not sufficient to
capture either the notion of that to which a theory is intended to
apply, or the notion of that to which it is thought to succeed in applying. But even if we set these failings aside, we find that the set-
theoretic conception faces yet another problem: that of providing an
account of progressive theory succession.

5. THE PROBLEM OF PROGRESS

The central notion in the set-theoretic view of how one theory might
be conceived to be superior to another is that of *reduction*, and the
idea it is intended to capture is, similar to the logical empiricist
notion of progress treated in Chapter 4, that a superior reducing

theory T' should be able to explain all that an inferior reduced theory T is capable of explaining, and more besides.[20] The reduction relation itself is essentially conceived to be a one-many relation (i.e. the converse of a function) from a subset of the set M_{pp} of partial possible models of the reduced theory T into the set M'_{pp} of partial possible models of the reducing theory T'. More particularly, it should be a relation from the set I of intended applications of T to the set I' of intended applications of T'. Furthermore, the models x and x' (elements of M_{pp} and M'_{pp} respectively) which correspond by virtue of this one-many relation are, in some sense, to be identical. According to Sneed, the fact that to each x there may correspond more than one x' suggests that T' is more complete, precise, and provides the means for making more distinctions than does T. Also, the reduction relation is to be such that if x' is a model of the predicate characterising T', then x is a model of the predicate characterising T.

One example where the sort of reduction sketched above is to obtain involves rigid body mechanics (T) and particle mechanics (T'), when each is defined as a set-theoretical predicate. Thus, for example, as regards the above requirement concerning identity, Sneed suggests that "it is possible to regard a rigid body and the particles that compose it as being, in some sense, the same thing."[21] But if we consider the relevant axiomatisations from which Sneed is working, namely, those given by Adams, we see that a rigid body and a particle are not (partial possible) models of their respective theories, but are members of members of such models. It is to be granted, however, that since on Adams' axiomatisation the first axiom of rigid body mechanics states that the first five (of seven) elements of a system (ordered n-tuple) of rigid body mechanics themselves constitute a system of particle mechanics, every model of the former theory is a model of the latter. In this case then we do have a relation of identity, though it is not one-many, but one-one, and does not involve intended applications, but models.

---

[20] For the notion of reduction given in the present paragraph, cf. Adams (1959), pp. 256ff., Sneed (1971), pp. 216ff., (1976), pp. 135ff., and Stegmüller (1973), pp. 127ff.

[21] Sneed (1971), p. 219.

Problems arise at this point, however, even if we simply follow the above line of reasoning, for it suggests that the set of models of a reduced theory is to constitute a subset of that of the reducing theory (and thus we might have a one-many relation between these sets). But this sort of relation would obtain in any case where the set of models of a predicate defined by certain axioms is compared with the set of models of a predicate whose definition is given by some but not all of those same axioms. Here the axioms of the reducing theory would be derivable from those of the reduced theory – not vice versa;[22] and 'progress' would result simply from the creation of new theories having fewer axioms than the old. Furthermore, far from being the more precise, as Sneed suggests, in this case the 'reducing' theory would be the less precise, in that fewer conditions would have to be met in order for a state of affairs to constitute one of its models.

However, as mentioned above, the reduction relation is not necessarily to obtain between M and M', but between I and I', and it is to be such that if x' is an element of M', then the x which corresponds to it by virtue of (the converse of) the relation is to be an element of M. Of course the conceiving of the reduction relation as obtaining between sets of intended applications is going to mean that it cannot receive a satisfying formal treatment until the problem of delineating the intended domain is solved. But even if (following Stegmüller) we were to conceive of the relation as being from one set of partial possible models onto another, it still faces the obvious difficulty that neither it nor its converse has been defined.[23] In other words, given some partial possible model of some theory, we are not told how to determine whether it corresponds to some partial possible model of some other theory, and so in no case can we be sure that one theory reduces to another in the set-theoretic sense. What we might do though is take the fulfilment of the conditions given at the beginning of this section as being *sufficient* for saying that one theory reduces to another, in which case (setting aside Sneed's identity requirement) an explicit definition of the relation between the sets of partial possible models would not be necessary. But, as has been noted by

---

[22] For a similar point, see Mayr (1976), pp. 279ff.
[23] This point has also been raised in Kaiser (1979), p. 16.

others, this would allow the construction of a 'reducing' relation between almost any two theories.[24]

In a more general vein, even if the set-theoretic conception of reduction were able to accomplish all that it set out to accomplish, it would provide no conception of how successive theories may be incommensurable. We might say that two such theories were incommensurable if, when defined as set theoretical predicates, they had quite different models – i.e. if their respective models were in no sense the same; but then neither would be reducible to the other. And if it were concluded from this that successive theories are then *not* incommensurable, it might still be asked how it is that they often conflict with one another. If we look at the situation simply from the point of view of set theory, each of the theories should have its own set of models, and it should or should not be the case that one of these sets is a subset of the other, or that the two sets overlap. But it is not possible to construe the two theories as conflicting, for even if the axioms defining one of them were to include explicit negations of axioms defining the other, this would only mean that the theories had different models, i.e. that they would apply, if at all, to different things.

6. CONCLUDING REMARKS

From the point of view of the present study the axiomatisation of Newtonian particle mechanics via the definition of a set-theoretical predicate is in itself quite acceptable. At the same time however it is noted that there are other, equally acceptable, ways of expressing the definitions fundamental to classical particle mechanics. But what is here seen to be a basic problem for the set-theoretic view as expounded by Sneed and Stegmüller is its assumption that the models arrived at via the principle of abstraction from the axioms in the definition are to exist in the real world – that they are or can be that to which the theory is intended ultimately to apply. Here these models are seen rather to be abstractions of the same sort as is the geometrical space defined by the Euclidean postulates, or the ideal gas model as defined in the previous chapter. This view of the situation explains

---

[24] E.g. on p. 262 of his (1959) Adams himself makes a remark to this effect.

why "the distinction between [physics and pure mathematics] is not clear from a set-theoretical standpoint,"[25] and also why this approach faces problems whenever questions regarding the empirical aspects of science are raised.

Thus where the set-theoretic conception is unable to explain such features of science as the intended domain of application of a theory, and how a theory may be considered successful though it does not quite fit the data, the present conception, with its intuitive basis in the Gestalt Model, is able to do so. Though Thomas Kuhn has himself suggested that the approach of Sneed and Stegmüller succeeds in capturing important aspects of his own view of science, the above considerations indicate that in the case of progressive theory succession – with which the present study is primarily concerned – the set-theoretic notion of reduction is unable to account either for theory conflict or for incommensurability, while the present conception provides an explanation of both of these aspects of science.

In the next and concluding chapter a brief look will again be taken at Newtonian mechanics, as well as at the laws of motion of Kepler and Galileo, and a sketch will be given of the approach to their views suggested by the Perspectivist conception. The opportunity will also be taken in this context to compare the present approach with the alternatives treated above on those points where the differences are particularly salient.

---

[25] Suppes (1967), Ch. 2, p. 52.

CHAPTER 12

APPLICATION OF THE PERSPECTIVIST CONCEPTION TO THE VIEWS OF NEWTON, KEPLER AND GALILEO

1. NEWTON'S THEORY OF GRAVITATION

What is commonly termed 'Newton's theory of gravitation' is presented in his book *Mathematical Principles of Natural Philosophy* (1687) in the following way. Eight definitions are first given, including, for example, the definition of quantity of matter (mass) as the product of density and volume. These definitions afforded, Newton next presents three *axioms*, or *laws of motion*, which are:

1. Every body continues in its state of rest, or of uniform motion in a straight line, unless it is compelled to change that state by forces impressed on it.
2. Change of motion is proportional to the force impressed, and is made in the direction of the straight line in which the force is impressed.
3. The forces two bodies exert on each other are always equal and opposite in direction.

From these axioms six corollaries are derived which include, for example, an explanation of how a body should move when acted upon by two forces simultaneously. Following this begins Book I: The Motion of Bodies, and the first pages of this part of the treatise are devoted to a development of the infinitesimal calculus (in a geometrical form), which later serves as an aid to the determination of, for example, centripetal forces – i.e. forces directed to a single point.

From this basis then, Newton proceeds to expound his conception of the nature of the motion of bodies, often invoking as a working hypothesis that which, when generalised, has come to be called his *law of gravitation*. This law may be expressed as follows:

Any two bodies attract each other with a force which is proportional to the product of their masses and in inverse ratio to the square of the distance between them.

Though Newton does not himself present this law in his book, it is clear from his assumption of it in particular cases that it underlies the whole of his thinking with regard to the motion of bodies, and might just as well have been included as a fourth axiom at the basis of his system. In any case, for the sake of simplicity we may here say that Newton's three axioms or laws of motion, together with his law of gravitation, define the basis of what is commonly called Newton's *theory of gravitation*.

The question often arises in considerations of Newton's theory as to whether his axioms or laws of motion ought best to be taken as definitions (or a priori truths), or as empirical laws, and the conclusion usually drawn is that they can function in both sorts of ways.[1] This conclusion is actually suggested by Newton's own presentation, where he calls them axioms *or* laws of motion; and it is easily accommodated on the present view. Here these expressions may be seen on the one hand to be definitions or axioms which, taken together, delineate the abstract *model*, in its simplest form, which underlies Newton's theory. Functioning in this way, Newton's axioms also tell us, for example, how we are to compute the values of certain functions (sometimes with the aid of explicit lemmas) in the context of the model, where bodies are conceived as mass-points having no extension. We note also that these particular expressions may be replaced by others, and from the present point of view the legitimacy of such a replacement in this context depends on whether the new expressions succeed in delineating a model in which the function-values obtained by computation remain the same. On the other hand, the present view also suggests that these same expressions may be seen as *theoretical laws* (as discussed in Chapter 10), in which case they are not to be employed simply in the context of the model, but are rather to constitute the first step to be taken in applying the model to empirical reality.

With regard to this issue we see that on the set-theoretic conception Newton's three laws are treated as axioms, and all considerations are restricted to the nature of the relations that obtain between elements (particles) in the model. The empiricist and Popperian views, on the other hand, go in quite the other direction and see these

---

[1] Cf. e.g. Hanson (1958), pp. 97ff., and Ellis (1965), pp. 61–62.

laws as applying to empirical reality, ignoring altogether the model they serve to define. In this way then the Perspectivist conception may be seen as a sort of amalgam of these two views, for it recognises the correctness of applying these laws or axioms in both directions.

2. KEPLER'S LAWS OF PLANETARY MOTION

One of the major achievements of Newton's theory of gravitation is considered to be its ability to explain Kepler's three laws of planetary motion. (Kepler published his first two laws in 1609, and his third in 1619.) Here we will investigate how, on the present view, Newton's theory may be seen as doing this, as well as why Kepler's laws taken together do not constitute a scientific theory. Kepler's laws may be presented as follows:[2]

1. The orbit of each planet is an ellipse, with the sun at one focus.
2. Each planet moves in its orbit at such a speed that a line joining it to the sun sweeps over equal areas in equal times.
3. The square of the time required for one revolution of a planet about the sun is proportional to the cube of its mean distance from the sun.

The reason usually given for Kepler's laws, taken together, not ranking as a theory is their unequivocally empirical character – i.e. the fact that they are 'instantial' in the sense that they refer to the individual planets in the solar system and, unlike a theory, are capable of being tested more or less directly.[3] The present view supports a distinction along these lines, and in fact provides an explanation of it, viz., that, unlike Newton's theory, Kepler's laws are not integrally related to a model, or, as Campbell might say, they do not evince any analogy, as they must in order to attain the status of theory.[4]

---

[2] For the sake of readability none of the laws considered in the present chapter have been expressed as equations.

[3] For a distinction between laws and theories along this line, see Feyerabend (1962), p. 28n. We note that, as remarked in earlier chapters, Popper's conception of science does not afford the making of this distinction, and that Popper himself treats Kepler's laws as constituting a scientific theory: cf. e.g. Popper (1957a), p. 198.

[4] Nor do they provide a notion of *cause*, as may be expected of a theory; in this regard see below, pp. 144 and 180, and Dilworth (2007).

The way in which Newton's theory explains each of Kepler's laws is in principle the same for each law, so we might simply restrict ourselves to a consideration of how this transpires in the case of Kepler's first law. We note that on Newton's model, when only two masses (mass-points) are considered – one being appreciably greater than the other and taken as non-accelerating – if the smaller mass is in motion with respect to the larger it will describe a conic with the larger mass at a focus. And the nature of the conic – i.e. whether it be an ellipse or a parabola – will depend on the velocity with which the smaller mass is moving. In order to explain Kepler's first law, this situation obtaining in the abstract model must be applied, with the aid of correspondence rules, to the solar system. This is done by associating the sun with the non-accelerating mass, and an individual planet with the other mass; these conditions set, the planet's velocity is then independently determined by empirical means. Given the particular results of this determination, Newton's theory suggests that, other things being equal, the planet should describe a perfect ellipse about the sun, one focus of the ellipse being occupied by the sun itself. It is in this way then that Newton's theory serves to explain Kepler's first law.

In discussing the nature of the relation between Newton's gravitational theory and Kepler's laws, Duhem, and following him Popper, has suggested that Newton's theory actually *contradicts* Kepler's laws.[5] What they intend by this is that, as is readily accepted on the present view, when the masses of more than one planet are taken into account, as well as the gravitational force exerted by the planet on the sun, then Newton's theory will, with regard to Kepler's first law for example, not suggest that the planet in question should describe a perfect ellipse, but rather that it should describe some other curve, since its orbit will be perturbed by the effect of its own mass and those of the other planets. But their use of the term "contradiction" deserves comment, for if this term is used in its philosophically precise sense then one of the two contradicting assertions must be true, whereas in the Newton-Kepler case, when the masses of all the planets are taken into account it might well be that neither the situation suggested by Newton's theory nor that depicted by Kepler's laws

---

[5] Cf. Duhem (1906), p. 193, and Popper (1957a), pp. 200ff.

actually obtains. And a second point in this regard is that if we flatly assert that the two systems contradict one another, not only is the fact obscured that Newton's theory and Kepler's laws are entities of essentially different sorts, but we will be hard put to account for the sense in which Newton's system may be seen to explain Kepler's. (This last point will be pursued below in the context of Galileo's laws.)

The way in which Newton's theory serves to explain Kepler's first law has been outlined above. Now we will show in somewhat more detail how, on the present view, this comes about, and at the same time the way in which the two systems nevertheless conflict. The idea here is that the relation between Newton's theory and Kepler's law is essentially of the same sort as the relation between van der Waals' gas model and Boyle's law or the general gas law (note: not the ideal gas model), as presented in Chapter 10. And the key notion in this regard is that of reduction, as it was given in that earlier chapter. Thus we should say that, when the value of the mass parameter is taken to the limit zero in the case of the planets in the solar system other than the one with which we are presently concerned, and the gravitational force exerted by the planet on the sun is neglected, then Newton's theory suggests identical results to Kepler's first law, and in this way explains that law. But, as in the case of the volume of the molecules in a real gas, in applying Newton's theory in order to understand the motions of all the planets as actually observed, these parameters are given positive values. In this case then Newton's theory (with the aid of correspondence rules) and Kepler's first law respectively suggest different results from the employment of the same instruments (telescope etc.) in the plotting of the paths of the planets. And it is in this sense then that the two systems are incompatible.

3. GALILEO'S LAWS OF INERTIA AND FREE FALL

Galileo, in his *Dialogues Concerning Two New Sciences* (1638), gives two laws for the motion of bodies. The first to be presented here has two forms, one being that if the earth were a frictionless sphere, a body moving on it would continue in its motion forever. This law – Galileo's *law of inertia* – will here be treated in its second form however, which Galileo arrives at by a further process of idealisation; in this form it may be presented as follows:

If a body is set in motion along a perfectly smooth horizontal plane, it will continue to move in such a way as is uniform and perpetual provided that the plane has no limits.

The second of Galileo's laws of motion is his *law for free fall*, which may be expressed as:

In falling freely towards the earth a body experiences constant acceleration with respect to the surface of the earth.

Here we note that in order to fall freely a body must be in a vacuum.

We might first consider whether Galileo's two laws, together, constitute a theory. On the present view it seems that they do, for these laws, in their explicit nullification of empirical factors, clearly depict an idealised model.[6] From this point of view, an infinite frictionless plane, for example, would exist in the model or theory, and that to which it corresponds in the physical world would be some local portion of the earth's surface. Thus we should say here that in this respect Galileo's laws of motion more closely resemble Newton's gravitational theory than do Kepler's laws.

On the basis of Galileo's two laws of motion we obtain what may be called a third law, which is that any projectile in motion relative to the surface of the earth (horizontal plane), when the effect of air resistance is neglected, will describe a parabola (the limiting case of which is a straight line). Newton suggests that this same result can be obtained on his theory if the centre of force (the centre of mass of the earth) is considered as being infinitely distant.[7] This of course is a counterfactual assumption similar to that regarding the absence of perturbing masses as is required in deriving Kepler's laws from Newton's theory. Thus we see that Newton's theory is capable of explaining Galileo's results concerning the motion of projectiles in the same sort of way as it is able to explain Kepler's laws, a particular instance of the distance parameter here being given a limiting (infinite)

---

[6] For comments on Galileo's two laws in this regard see Krajewski (1977), pp. 18 and 26. (We note however that it may be the case that the depiction of an idealised model need not in itself suffice for an expression's being that of a theory, though this question will not be pursued here.)

[7] Cf. the scholium following Prop. X, Prob. V in Bk. I of Newton (1687).

value; and the method employed is again one of reduction, as this term is understood in the present study.

In discussing the nature of the relation between Newton's theory and Kepler's and Galileo's laws respectively, Popper has said:

Newton's theory unifies Galileo's and Kepler's. But far from being a mere conjunction of these two theories – which play the part of *explicanda* for Newton's – *it corrects them while explaining them*. ... Far from repeating its *explicandum*, the new theory contradicts it and corrects it.[8]

Once again, however, as in Chapter 6, we have an instance of Popper presenting a relatively acceptable description of what occurs in actual science, but one which cannot be captured on his own conception. On Popper's view, as has been treated in detail in the earlier chapters of this study, explanation consists in deductive subsumption. Here, however, he wants to apply this notion not only to a case where such subsumption does not obtain, but to one which, on his own way of thinking, involves contradiction. Popper wants to have his cake and eat it too, and even goes so far as to provide the icing with his further suggestion that Newton's theory *corrects* its predecessors – a notion nowhere treated in the actual presentation of his conception of how competing theories are related to one another.

On the present conception however it may be said that, while Newton's theory, in application, is perspectivally incompatible with the 'theory' lying behind Galileo's laws, and suggests results differing both from those expected on the basis of Galileo's laws and from those suggested by Kepler's laws, it provides a unified account of their respective domains: planetary and terrestrial motion. It places both domains within one conceptual perspective. Following the Gestalt Model we may see that these unified domains can be presented in a way which is independent of the theories or laws concerned: just as the domain of application of the 'duck' and 'rabbit' concepts may be described as a line on a piece of paper, the realm of application of the above views may be said to consist in the data resulting from measurements of position and velocity etc. of certain physical bodies. These data constitute a 'background of accepted fact' which, while not *absolute* in that its viability is based on certain empirical laws, is nevertheless *neutral* as regards the views under

---

[8] Popper (1957a), p. 202.

consideration. In this regard we might consider for example the data accumulated by Tycho Brahe on the basis of which Kepler discovered his three laws of planetary motion: these data are neutral with respect to both Kepler's laws and Newton's theory, and constitute a realm to which each can be applied – in this way the data might conceivably have played an independent role in the debate between them. Similarly, neutral data would of course be available for an evaluation of Galileo's and Newton's theories.

Now, on this view it is not necessary that any of the above laws or theories should take full account of all of the relevant data. With the passage of time after the determination of initial conditions we might expect that Newton's and Galileo's theories and Kepler's laws would be completely off in their predictions (or retrodictions). Nevertheless, those predictions concerning terrestrial and planetary motion made on the basis of Newton's theory give numerical values *closer* to the values obtained using instruments in actual measurements. On the basis of such evidence, in conjunction with the fact that Newton's theory provides a unified or simple account having a broader scope than either of its rivals, we may say that scientific progress has been made in moving from the laws of Kepler and Galileo to the theory of Newton.

But the superiority of Newton's theory in this respect should not mean the utter dismissal of its rivals. Kepler's laws, for example, afford a description of planetary motion which is sufficiently accurate for many purposes, and which has the advantage of being conceptually more manageable than the account provided by Newton. Nor does this superiority suggest that Galileo's and Kepler's laws are false, while Newton's theory is true, or more true. Rather, it implies that, of the three, Newton's theory is the best *applicable* to the domain of terrestrial and planetary motion.

Thus at the same time as affording a conception of theory conflict as perspectival incompatibility, and incommensurability as shift of conceptual perspective, the present conception avoids relativism through the provision of a realistic notion of scientific progress as depending on the results of measurement. It may be kept in mind though, that while the above might function as a criterion as to what should constitute progress in science, it need be no guide as to what in any broader realm should constitute epistemological advance.

APPENDIX I

ON THEORETICAL TERMS

1. THE LOGICAL EMPIRICIST NOTION OF THEORETICAL TERM

Though the question of the nature of theoretical terms has been the subject of much recent discussion in the philosophy of science, little of that discussion has taken account of what gave rise to the issue in the first place. The notion of theoretical term first became of relevance to the philosophy of science when it was realised that a number of scientific expressions would apparently not be meaningful according to a widely held (logical empiricist) conception of meaningfulness. On this conception, in order for an utterance to be meaningful, either its non-logical constituents must themselves refer to entities which are directly observable, or they must be definable in other terms, the referents of which are directly observable. Thus such scientific terms as "atom" and "electron," and "temperature" and "magnetic," would all appear to be meaningless, since their referents are seemingly unobservable.

This state of affairs constitutes a major problem for the logical empiricists' conception of meaningfulness (or, alternatively, for their conception of science). And it is in the context of this problem that the notion of theoretical term obtained its first characterisation, namely, as being that of any non-logical scientific term which does not have a directly observable referent.

Three implications of the above are particularly noteworthy: first, the introduction of the notion of theoretical term into the philosophy of science was motivated by considerations based on a particular view of science in which its linguistic aspects are emphasised;[1] second, that theoretical terms should differ in an essential way from observational terms is not a requirement of the empiricist view, but constitutes rather a problem for it; and third, the original characterisation of the notion of theoretical term does not require that such a term actually occur in a theory.

---

[1] See below, pp. 258–266.

2. CARNAP AND 'THE METHODOLOGICAL CHARACTER OF THEORETICAL CONCEPTS'

Various attempts have been made to overcome the logical empiricists' problem of 'theoretical' terms. However, none of them has been altogether successful, a fact which may account for a shifting of interest from the meaningful/meaningless distinction to the observational/theoretical distinction, with the concomitant implicit suggestion that the latter distinction is in some way supported by the empiricist view of science.

This shift of interest may perhaps be most clearly seen in the opening lines of Carnap's paper, 'The Methodological Character of Theoretical Concepts,' where he says: "In discussions on the methodology of science, it is customary and useful to divide the language of science into two parts, the observation language and the theoretical language."[2] No mention is made however of the fact that the suggested division is customary due to the existence of the empiricists' problem of theoretical terms, nor that it is useful from an empiricist point of view only in the sense that some such division must be recognised in order to deal with the problem. Though Carnap does go on in his paper to grapple with the problem itself (assuming that theoretical terms are in fact meaningful, and further weakening the empiricist criterion of meaningfulness in order to accommodate them), his approach has nevertheless left some philosophers with the impression that his aim is not so much to save some sort of empiricist criterion of meaningfulness as to maintain a clear distinction between observational and theoretical terms.

3. 'PUTNAM'S CHALLENGE'

Hilary Putnam, in his paper 'What Theories are Not,' presents a critique of the 'received view' of scientific theories, in which they are conceived of "as 'partially interpreted calculi' in which only the 'observation terms' are 'directly interpreted.'"[3] But Putnam does not distinguish this view, arrived at by logical empiricists in their attempt

---

[2] Carnap (1956), p. 38.
[3] Putnam (1962), p. 240; next quote, ibid., p. 241.

to handle the fact that there do exist terms in science which do not have observable referents, from the historically antecedent logical empiricist view itself, with its observability requirement for meaningfulness. Thus, for example, he says:

[I]t should be noted that the dichotomy under discussion was intended [e.g. by Carnap] as an explicative and not merely a stipulative one. That is, the words "observational" and "theoretical" are not having arbitrary new meanings bestowed upon them; rather, pre-existing uses of these words (especially in the philosophy of science) are presumably being sharpened and made clear.

As has been pointed out above, however, Carnap's interest was not to explicate the notions of theoretical and observational, but to show how terms in science which do not refer to observables can nevertheless be construed as meaningful on a (modified) empiricist view.[4]

We might then keep in mind that what Putnam refers to as the 'received view,' with its presupposition of the possibility of drawing a clear distinction between observational and 'theoretical' terms, was arrived at by logical empiricists in attempts to retain the essence of their conception of meaningfulness in the face of certain unwelcome facts concerning the nature of science. Thus the work of Putnam (and Achinstein) may perhaps most profitably be seen in the present context as suggesting that, independently of any particular philosophical leaning, philosophers of science might further investigate the observational/theoretical distinction, and consider whether or how such a distinction ought to be made. It is in this positive spirit that Wolfgang Stegmüller views Putnam's paper, where he says:

Putnam's reproach to the effect that no one has attempted to clear up the specific role played by theoretical terms within a theory we will call *Putnam's challenge*. Its substance is contained in the following sentence: "A theoretical term, properly so called, is one which comes from a scientific *theory* (and the almost untouched problem, in thirty years of writing about 'theoretical terms,' is what is *really* distinctive about such terms)."[5]

---

[4] This point has also been missed by Peter Achinstein in his treatment of the subject. Cf. e.g. Achinstein (1965), p. 196, and (1968), p. 158.

[5] Stegmüller (1976), p. 27; Putnam (1962), p. 243; also cited in Sneed (1971), p. 34.

4. SNEED'S NOTION OF THEORETICAL TERM

J. D. Sneed, in his book *The Logical Structure of Mathematical Physics*, introduces a notion of theoretical term which Stegmüller points to as being the first attempt made to answer Putnam's challenge. Sneed introduces his notion of 'theoretical' in the context of his own conception of science, in which scientific theories are conceived to be set-theoretical *n*-tuples. In this context he suggests a theoretical term to be the numerical value of a parameter whose measurement presupposes that the theory in which the parameter occurs has already been successfully applied.[6] Since the same parameter may occur in different theories, we see that for Sneed a particular term may be theoretical with respect to one theory while not being theoretical with respect to some other.

Has Sneed succeeded in answering 'Putnam's challenge' through providing a characterisation of theoretical terms which reveals what is really distinctive about them? In this regard it may first be noted that Sneed's characterisation is limited to measurable parameters. Thus, for example, where the term "electron" – which is to occur in a theory and refer to an unobservable – should constitute a paradigmatic instance of a theoretical term for Putnam, for Sneed it would count neither as a theoretical nor as a non-theoretical term, since both sorts of term are conceived by him to be numerical values.

One might accept Sneed's exclusion of such terms as "electron" from consideration, were he to support this move with convincing argument, or were his positive characterisation to deepen our understanding of the situation so that their exclusion became self-evident. But Sneed provides no supporting argument; and – a second point regarding his characterisation – it by no means clarifies the role of theoretical terms in science, but leads immediately to an infinite regress, thereby making us wonder why such terms should exist in science at all.

A third point is that, while Sneed's characterisation treats of theoretical terms, it does not treat of empirical terms. In fact, on Sneed's way of thinking, *all* mensural functions are theoretical with respect to at least one theory (namely, that in which they are implicitly defined), and so science is to contain no purely empirical or

---

[6] For greater detail, see *this volume*, pp. 109–113.

non-theoretical terms. In conjunction with his characterisation, this has the effect of denying science an empirical basis. This is not only a surprising result, but one which is made all the more difficult to accept by the fact that Sneed does nothing by way of supporting it.

A successful answer to 'Putnam's challenge' regarding what is truly distinctive about theoretical terms in science should do more than merely posit a criterion for singling them out, especially when the criterion is as problematic as that suggested by Sneed. What more may be expected is a clarification of why such terms should exist at all – what role they play in the scientific enterprise. And neither the logical empiricists nor Sneed have succeeded in explaining what that role is.

5. MEASUREMENT AND THE EMPIRICAL BASIS OF SCIENCE

A feature common to both Sneed's and the logical empiricists' characterisation of theoretical terms is the treatment of mensural terms as being theoretical. For Sneed, all mensural functions are theoretical with respect to the 'theories' constituting their definitions; and for empiricists such as Carnap and Hempel, mensural terms such as "temperature" are theoretical because their referents cannot be observed directly.[7]

In the case of Sneed, as mentioned above, science contains no purely empirical terms; and in the case of the empiricists, the situation is much the same. Assuming mensural terms to be theoretical, expressions which are normally accepted in science as being of empirical laws (e.g. Boyle's law, Ohm's law) must be conceived as being of theories, due to their express inclusion of such terms. Thus, as in the case of Sneed, virtually all scientific notions become 'theoretical'; and what the empiricists recognise as empirical terms find essentially no employment in science at all. Moreover, this in turn suggests that actual science contains few if any truly empirical laws.

The fact that scientists themselves employ the term "empirical" in referring to certain numerical laws thus leads to a questioning of that aspect of the logical empiricist empirical/theoretical distinction which sees mensural terms as falling on the theoretical side of the

---

[7] Cf. Carnap (1956), p. 48; Hempel (1965), p. 185.

demarcation. What is more, if we were to follow the empiricists in considering subjective sensations and perceptions as constituting the data of an empirical enquiry, then the natural sciences simply could not be seen as being empirical enquiries. Since they are so seen, the question becomes: in which sense then is science empirical, or, what constitutes what we consider to be the empirical aspect of science?

This latter question has in fact already received an answer which, from the present point of view, is quite acceptable. As has been suggested by N. R. Campbell, the empirical aspect of science consists of the network of known empirical laws. Campbell considers (the expression of) an empirical law to be one which has not been derived from theory – but this is a negative characterisation. So let us suggest here, more positively, that an empirical law is one the existence of which has been established on the basis of the results of particular operations, most often, operations of *measurement*; and that the term "empirical" as it is used in the exact sciences means essentially nothing other than 'mensural,' i.e. 'having to do with measurement.'

Consideration of the above suggestion, however, may lead to the raising of another question, namely: does the characterisation of that which is empirical in science as being mensural allow the making of a clear demarcation between the empirical and the theoretical? Or, in other words, what then is the essential difference between empirical (mensural) laws and scientific theories (for is it not the case that measurements themselves presuppose theories of measurement)?

Unfortunately, those who suggest that measurements necessarily involve an element of theory have devoted no attention to the possibility that what measurements necessarily involve are not theories, but *laws*. The next section of this appendix is devoted to an analysis of some of these views; but in what follows of the present section the above-mentioned example of temperature will be considered, so as to illustrate the way in which mensural results are in fact determined by the obtaining of certain laws, while the particular form these expressions take may itself be suggested by scientific theory.[8]

We today take alterations in temperature, as manifest in thermometer readings, to be the result of changes in the rate of motion of

---

[8] For a treatment of temperature measurement comparing it with other kinds of measurement, see Ellis (1966), Ch. VI.

submicroscopic particles. This conception of temperature is based on the kinetic theory of matter, and is in this way theory-dependent. However, thermometers were used to measure temperature long before the kinetic theory was generally accepted, for example in the establishment of the general gas law, which states that the pressure of a given volume of gas varies with its temperature. And, if the kinetic theory had never been proposed, or if another theory had been advanced instead, this would have had no effect on the facts requiring explanation, namely, that under certain specified conditions, thermometers gave certain particular readings.

The general point which is being suggested here is that measuring instruments, in being constructed to accord with intersubjective and non-changing physical standards, when used in a way which is clearly specified and appropriate to their design, provide numerical results which constitute the data-base of the exact sciences. The fact that such results are replicable suggests that they are the manifestation of regularities of nature, that is, of natural laws. Though it requires ingenuity to discover these laws amongst the data, or through the performance of new experiments with measuring instruments, the same measuring operations performed under the same conditions will always produce the same results, independently of the theory employed to interpret or explain those results. In the case of temperature then, independently of the mechanism responsible for thermometers working in the way that they do, we can and do rely on their readings in the establishment of such physical laws as the general gas law.

Once one has adopted a particular theory, however, the kinds of measurements one makes, and even the instruments one uses in making them, may change. Having adopted the kinetic theory of matter for example, and thus assuming that all things consist of minute particles in various states of motion, it is clear that the motion of the particles taken to constitute a liquid is restricted in a way that the motion of those constituting a gas is not. Since, on the theory, temperature is nothing other than the very motion of such particles, a gas rather than a liquid would be a more reasonable choice of thermometric substance. In this way then, one might have set about constructing and using a gas thermometer, had such a device not been developed before. And the results obtained using such an instrument would have differed somewhat from those obtained using a liquid thermometer, as would the expression of the laws dependent on those results.

Thus we see how, on the one hand, the particular results obtained using measuring instruments are dependent upon natural law, while, on the other, the manner in which a measurement is made can be largely determined by scientific theory. One is free to construct measuring devices as one will, according to some theory one has adopted, or in order to test some hypothesis. But what one cannot do is alter the results of the measuring operation itself. These results, or the empirical laws in which they are systematised, constitute the scientific facts, which together comprise the empirical aspect of science.

6. ON THE 'THEORY-LADENNESS' OF ALL SCIENTIFIC CONCEPTS

Largely as a result of the failure of logical empiricism to substantiate the conception of science as having a purely phenomenal basis, a number of philosophers have claimed that all of the terms employed in science involve an element of theory. However, these philosophers have not considered extending the meaning of "empirical" to include 'that which can be measured,' nor the possibility of there being a difference in principle between laws and theories. As a consequence, they have devoted no time to a close examination of the actual presuppositions made in the performance of measurements, and have furthermore been inclined to treat as theoretical that which is founded on general laws.

Karl Popper, for example, has said: "For even ordinary singular statements are always *interpretations of 'the facts' in the light of theories*. (And the same holds even for 'the facts' of the case. They contain *universals*; and universals always entail a *law-like* behaviour.)"[9] Not only does Popper not distinguish between laws and theories, but he conceives of both as simply being universal statements, and gives no consideration to the role played by measurement in the establishment of laws as they are actually presented in science.

---

[9] Popper (1959), p. 423.

The distinction between laws and theories, and the role of measurement in science, is also overlooked by N. R. Hanson in those parts of *Patterns of Discovery* where he suggests that scientific terms are, on the whole, theory-laden. In the context of an analysis of the nature of causality, Hanson says: "Galileo often studied the moon. It is pitted with holes and discontinuities; but to say of these that they are craters ... is to infuse theoretical astronomy into one's observations." And he continues: "Sketches of the moon's surface would just be sketches of a pitted, pockmarked sphere; but Galileo saw craters."[10] Though in his list of terms which he takes to be theory-laden in a way similar to "crater" Hanson includes such mensural terms as "pressure," "temperature" and "volume," and "force," "momentum" and "velocity,"[11] he does not work through a real scientific example in this context involving any of them, and so leaves it unclear as to why we should see the notions behind such terms as resting on theories rather than laws.

The question of the 'theory-ladenness' of mensural notions is treated directly, however, by Paul Feyerabend. In 'Problems of Empiricism' he says: "Taken by themselves the indications of instruments do not mean anything unless we possess a *theory* which teaches us what situations we are to expect in the world, and which guarantees that there exists a reliable correlation between the indications of the instrument and such a particular situation."[12] He goes on to suggest that if a certain theory is replaced by another with a different ontology, then we may have to revise the interpretation of all our measurements; and he provides the following example: "[A]ccording to the phlogiston theory, measurements of weight before and after combustion are measurements of the amount of phlogiston added or lost in the process. Today we must give a completely different interpretation of the results of these measurements."

As has been suggested by H. R. Post, however, "The example given by Feyerabend goes no further than making the point that the 'interpretation' of a measurement is indeed theory-dependent .... This does not provide an example of incommensurable descriptions of measurement itself."[13] And, as regards Feyerabend's emphasis upon

---

[10] Hanson (1958), p. 56.
[11] Ibid., pp. 61 and 65.
[12] Feyerabend (1965c), pp. 36–37; next quote, ibid., p. 37.
[13] Post (1971), p. 252. Next quote, ibid., p. 253; cf. also *this volume*, pp. 129–130.

the shift of ontology which is to accompany theory change, Post goes on to note that: "the statement 'the pointer coincides with the mark' neither asserts nor presupposes the existence of atoms or of a continuous type of matter; such a neutral observation statement may thus well be shared by or critically decide between two theories differing in these fundamental ontological presuppositions."

A point not taken up by Post, on the other hand, is that not only might a singular statement about an instrument reading be independent of the theories to which it is relevant, but, as is suggested by the considerations of the previous section, empirical laws themselves can and must be formulable and testable without reference to any theory claiming to explain them. As expressed by Ernest Nagel: "Despite what appears to be the complete absorption of an experimental law into a given theory, so that the special technical language of the theory may even be employed in stating the law, the law must be intelligible (and must be capable of being established) without reference to the meanings associated with it because of its being explained by that theory."[14] Nagel provides a number of examples from the history of science in support of his claim, including e.g. that, with respect to Millikan's oil-drop experiment,

the truth of the experimental law that Millikan helped to establish (namely, that all electric charges are integral multiples of a certain elementary charge) is not contingent upon the fate of the [electron] theory; and, provided the direct observational evidence for it continues to confirm the law, it may outlive a long series of theories that may be accepted in the future as explanations for it.

It might still be objected to the above that, though certain scientific notions may be independent of particular theories, it has not been shown that such notions are independent of *all* theories. Though this stronger view is not being argued for here, its denial might well lead to awkward relativistic consequences. What is being argued for here, on the other hand, is that a distinction may be made in principle between empirical laws, characterised by their intimate relation with the results of mensural operations, and scientific theories, which may or may not be characterised by their containing 'theoretical notions,' a question to which we now turn.

---

[14] Nagel (1961), p. 87; next quote, ibid., p. 88.

7. CAMPBELL, 'HYPOTHETICAL IDEAS' AND THE IMPORTANCE OF ANALOGY

According to Campbell, the fundamental purpose of scientific theories is the explanation of laws. Laws are conceived to relate measurable concepts, while theories each consist of two parts: an 'hypothesis,' which contains propositions referring to hypothetical *ideas*; and a 'dictionary,' which links the hypothetical ideas in the hypothesis to the measurable concepts in the laws the theory explains. It may often be that not all of the hypothetical ideas of a theory can be so linked, in which case the hypothesis, unlike a law, will be truly hypothetical in that it is incapable of being directly tested by experiment.

Campbell suggests that "the hypothetical ideas of most of the important theories of physics ... are mathematical constants and variables;"[15] and he provides further clarification of his distinction between ideas and concepts where he says:

[T]he only numerical values which can be attributed to concepts as the result of experiment are finite values; the hypothesis on the other hand involves, by the mention of differential coefficients, the attribution of infinitesimal values to the hypothetical ideas. [T]here must consequently be a distinction between the hypothetical ideas and the concepts.

Not all hypothetical ideas need be mathematical however, and some of them may be distinguished from concepts by, for example, their referents' simply being too small or too distant in the past.[16] As has been remarked by Gerd Buchdahl,[17] Campbell's hypothetical ideas lack the direct instrumental relation possessed by concepts for a number of reasons, and Campbell does not always distinguish among them. Thus a general characterisation of what Campbell means by 'hypothetical ideas' would be essentially negative, viz., that they are notions used in referring to that which, for one reason or another, cannot be directly measured.

As will be suggested by what follows, this characterisation is in fact too broad, for, as Campbell himself realises, it leads to the classification of certain expressions which one would intuitively take to be of laws as being of theories. But Campbell does not put the whole

---

[15] Campbell (1920), p. 123; next quote, ibid., p. 142.

[16] Cf. Campbell (1921), pp. 96–97.

[17] Buchdahl (1964), pp. 155–156.

weight of his distinction between laws and theories on the latters'
containing hypothetical ideas while the former do not. He suggests
further that "in order that a theory may be valuable it must have a
second characteristic; it must display an analogy."[18] And earlier he
says in summary: "This analogy is essential to and inseparable from
the theory and is not merely an aid to its formulation. Herein lies the
difference between a law and a theory, a difference which is of the
first importance."

Campbell takes the analogy displayed by a theory to be to known
laws – a requirement which is perhaps too stringent – and distinguishes
between 'mathematical' (physical) theories, in which the analogy is to
the laws the theory explains, and 'mechanical' theories, in which the
analogy is to other laws. The importance of the analogy for Campbell
is underlined by the fact that he considers the hypothesis actually to
be based on it.[19] In this way the analogy acts, on the one hand, as a
constraint on the form of the hypothesis, while at the same time it constitutes a source of insight as to how the hypothesis might be further
developed.[20] But Campbell nowhere explains whether or how such
analogies must give rise to non-measurable ideas in the hypothesis.

Granting that in the case of at least certain theories, such as the
kinetic theory of gases, Campbell is justified in emphasising the
importance of analogy, an investigation of the relation between analogies and hypothetical ideas might prove valuable for our attempt
to provide a positive characterisation of theoretical terms, and clarify
the role they play in the scientific explanation of laws. In keeping
with such authors as Mary Hesse and Rom Harré however, we shall
not follow Campbell in emphasising the notion 'analogy,' but direct
our considerations rather to that which is constructed on the basis of
the analogy, namely, the *model*.

8. SOURCE VS. SUBJECT OF A MODEL

That model construction plays a fundamental role in scientific theorising has been indicated by a number of authors, including Fritz

---

[18] Campbell (1920), p. 129; next quote, ibid., p. 119.
[19] Ibid., p. 133.
[20] In this regard, cf. Hesse (1966), pp. 4 and 8ff.

Machlup (1952), E. H. Hutten (1954), and Reinhold Fürth (1969), as well as by Leszek Nowak, who takes the basic theoretical activities of a physicist to be "the construction of models of different sorts: the model of the perfectly rigid body, the model of the homogenous cosmos, the model of the perfect gas, etc."[21] For Nowak, the laws which are valid for such models may themselves be called *idealisational*. With time, these laws are gradually modified so as to provide a more realistic description of the actual phenomenon in question, a process which Nowak calls *concretisation*, and which goes hand in hand with the construction of more and more realistic models of the phenomenon. van der Waals' law, for example, is in this way a concretisation of the law valid for perfect gases, for, by taking account of factors neglected by the latter, namely the volume of the molecules of the particular gas, as well as certain forces acting between the molecules, van der Waals' law comes nearer to depicting the way real gases behave.[22]

Rom Harré, in discussing the nature of models employed in scientific theorising, distinguishes between *idealisations* and *abstractions*, the latter being models similar to those for which Nowak takes idealisational laws to be valid. For Harré, an idealisation is a model the properties of which are in some way more perfect than those of its 'source-subject,' while an abstraction is a model having fewer properties than its source subject,[23] the *source* of a model being that in analogy to which the model is constructed, and its subject here being taken as that to which the model is applied.

Nowak however, in his treatment of idealisation, overlooks the role played by the model's source, and so considers explanation to consist simply in the alteration of a law which is originally valid for the model itself in such a way that it becomes valid for the model's subject. Thus, for example, he considers the explanation of the behaviour of real gases to consist in the injection of notions representing the secondary factors of molecular volume and action at a distance into the equation expressing the ideal gas law. But this way of seeing the situation provides no rationale for the introduction of the notion of molecules, nor for the existence of the sophisticated model for which van der Waals' law is valid. While the ideal gas law may be

---

[21] Nowak (1979), p. 285.
[22] Cf. *this volume*, pp. 93–94 and 101–102.
[23] Harré (1970a), pp. 41–42.

seen as depicting a model which is more abstract than that depicted by van der Waals' law, neither law by itself depicts a model providing a representation of the substance or mechanism which is to be responsible for the law's taking the form that it does. Explanation does not consist in the gradual 'concretisation' of an abstract law until it becomes realistic; it consists in the derivation of a law, which itself applies relatively well to certain experimental situations, from a model or theory depicting the nature of its causal basis.

Following the present distinction between source and subject, we should say that the models constituting the core of the kinetic theory of gases are constructed in analogy to the notion of physical balls bouncing off one another – this is the source of such models, while their subject is the gas laws. The explanation of Boyle's law – a law which is itself sufficiently 'factual' to be applied directly to concrete situations – is obtained by deriving it from the ideal gas model, which depicts measurable pressure as the direct result of the impact of molecules. It would also seem, however, that both of Harré's notions of idealisation and abstraction should apply to the kinetic theory, in spite of its source differing from its subject. It is an idealisation in that its constituents interact in rigidly defined ways, and it is an abstraction since it does not employ notions such as colour or volume, as may be present in its source, nor temperature, which is present in its subject.

Mary Hesse, in her development of Campbell's view, explains the abstraction of the model from its source in terms of negative, positive and neutral analogy.[24] The negative analogy is that aspect of the source which is not contained in the model – in the above example involving the kinetic theory, part of the negative analogy would thus be the colour of the physical balls. The positive analogy is that aspect of the source which is retained in the model, e.g. the sphericity of the balls, and the fact that they have mass and can collide with one another. And the neutral analogy is that which is present in the source and which may prove a fruitful addition to the model, depending on whether its inclusion seems warranted on the basis of empirical tests. An example of the incorporation of such a neutral analogy is van der Waals' treatment of the molecules in his gas model as having non-negligible volume.

---

[24] Cf. Hesse (1966), pp. 8ff.

9. MODELS AND THEORETICAL TERMS

Against the background of what has been said above, we should say that the *subject* of a theoretical model consists in the facts represented by the data obtained by measurement in experimental situations, or that it consists in the scientific laws based on such data. These data are objective in that they are themselves based on physical standards or *paradigms*, and are obtained by the performance of clearly specified operations susceptible of intersubjective scrutiny.[25] What may also be noted in the present context is that these data or laws relate *properties*, such as pressure, temperature, or current flow, and do not make explicit reference to the bearers of such properties.

In seeking to understand why the laws of science take the form that they do, the scientist attempts to conceive what the reality underlying them must be like. He thus constructs an idealised model, which has as its *source* that which he feels he does understand – often things with which he is familiar in his everyday life. The model should be constructed so as to depict a physically possible, albeit idealised, reality, whose existence would naturally manifest itself in the laws requiring explanation.

We see here then that a scientific model represents an ontology the nature of which may be taken as being responsible for the epistemology we associate with scientific laws. And, as is implicit in what has been said above, the very fact that an explanation of such laws should be felt necessary suggests that that aspect of reality which is responsible for them is not open to direct inspection. Thus, if we consider models which depict these (perhaps temporarily) hidden aspects of reality as constituting the essence of scientific theories, we can characterise theoretical terms as terms used in referring to those entities in the real world (should they exist) as are depicted in such models.

10. ON THE EMPIRICAL AND THEORETICAL ASPECTS OF SCIENCE

The question of the nature of theoretical terms, as posed in 'Putnam's challenge,' is implicitly a question concerning the nature of scientific

---

[25] In this latter regard, see Agazzi (1977a), pp. 162–164 and 167–168.

theory itself, for it is only when one has a clear conception of what a theory is, that the nature of theoretical terms can be explained. And an understanding of the nature of scientific theory requires, in its turn, an appreciation of the nature of that which is considered in science not to be theory, but fact, i.e., the empirical aspect of science.

That the notion of theoretical *term* should be thought to require explication in the philosophy of science, as has been mentioned above, is a consequence of the linguistic approach taken to science by logical empiricism, with its observability criterion of meaningfulness. Thus the fact that Sneed should feel his own view of science to be threatened by a problem of theoretical terms appears somewhat curious, since his basis is not linguistic, but set-theoretical. However, with regard to Sneed and the empiricists, what may be emphasised here is that neither of them have succeeded in providing a conception of science which does justice to both its empirical and its theoretical aspects. In the case of Sneed, no conception is provided of empirical laws; and in the case of the empiricists, no explanation is given of the nature of scientific theory.

In the present appendix the attempt has been made to clarify the nature of theoretical notions through the presentation of an overview which includes both the theoretical and the empirical aspects of science. Where the empirical aspect is seen to include experimental laws and be based on operations of measurement, the theoretical aspect is seen to consist in the construction of idealised models capable of explaining, in terms of the action of physical entities, the relations posited to hold by the experimental laws. In this way, as suggested above, theoretical terms may themselves be explained as terms used in referring to the entities depicted as existing in such idealised models.

REPLY TO CRITICISM OF THE FIRST EDITION

1. LOGICAL EMPIRICISM AND POPPERIANISM

Much confusion has been evinced on the part of my reviewers as regards the nature of the Deductive Model. The Deductive Model is named and presented explicitly for the first time in this book, but it has been used implicitly on countless occasions in the philosophy of science. And while its logical form is identical to that of the covering-law model of deductive-nomological explanation (the D-N model), its range of application is much broader; furthermore, where the purpose of the D-N model is to depict the structure of scientific explanation and prediction, the purpose of the Deductive Model – as it is being employed in this book – is to show that both the logical-empiricist and Popperian conceptions of science are *dependent* on this structure. To this end, I reconstruct the whole of their respective views in terms of it. As I say in the text:

This reconstruction is intended to show in detail what the logical empiricist and Popperian views consist in, and in so doing demonstrate how they are in fact conceptually dependent on the Deductive Model, which thereby determines both their capabilities and their limitations.[1]

According to *Peter Gärdenfors*, in the first part of the book I presuppose that both the logical empiricists' and Popper's and Lakatos' understandings of science can be structured in accordance with 'the deductive model.'[2] Assuming that by 'the deductive model' Gärdenfors means the Deductive Model, I do not *presuppose* that these philosophers' conceptions of science can be structured in accordance with it, I so *structure* them! As John Blackmore says, "This book provides an extremely clear *description and critique* of the best known contemporary versions of philosophy of science."

---

[1] *This volume*, pp. 1–2.
[2] Gärdenfors (1982), p. 44; next quote Blackmore (1984), emphasis added.

Gärdenfors continues that while my criticism of the empiricists and Popper is essentially correct, it contains no novelty.[3] That he should say so however suggests either that he has not read my criticisms or is unfamiliar with those that have been made by others. I believe that a number of the criticisms that I make are quite original; and where they are not, I cite the source of the criticism in question. My own criticisms include that the logical empiricist and Popperian views are formally identical (pp. 11ff.); that it is impossible for Popper to distinguish science from non-science on the basis of falsifiability once he claims, as he does in criticism of the empiricists, that no empirical statement is verifiable (p. 14); that his notion of corroboration is formally identical to the logical empiricist notion of confirmation, and thereby commits him to induction (pp. 17–18) – for which he also criticises the empiricists; and that he provides no notion of content applicable to contradicting 'theories' (p. 34). To this I would like to add that, so far as I know, not only are all of these criticisms original, but, as Gärdenfors admits, they are also *correct*. Furthermore, and most important, I reconstruct the *whole* of logical empiricism (including providing the empiricists with a formal conception of scientific progress)[4] *and* Popperianism in terms of the Deductive Model, which, among other things, has the implication that Popper's philosophy of science is not so different from that of the empiricists as he would have us believe. (Cf. Blackmore: "Dilworth is particularly acute in pointing out the similarities in the logical empiricist and Popperian methodologies of science.") Gärdenfors misses not only the novelty in my doing this, but the very fact *that* I have done it.

Aant Elzinga states that the first part of the book consists of "a critique of the covering law model of explication."[5] Taking him to mean the covering-law model of *explanation*, his saying this mistakenly suggests not only that the first part of the book has to do with the covering-law model rather than the Deductive Model, but also that the object of my criticism is the model rather than the empiricist and Popperian conceptions of science.

---

[3] This claim has also been made in a review of the third edition of the book, and has received mainly the same response: cf. Dilworth (2007), p. 273.
[4] *This volume*, pp. 19–21; parenthetical quote following, Blackmore (1984).
[5] Elzinga (1981–1982), p. 267.

Douglas Shrader, for his part, also conflates the Deductive Model and the D-N model, stating that most of the shortcomings of the Deductive Model are already well known to the majority of people in the field, and that a person well-versed in the literature could, without serious loss, omit or skim the first six chapters of the book.[6] Thus, like Gärdenfors, he misses what is original in my individual criticisms, and fails to realise that what "people in the field" may be familiar with are certain criticisms of the D-N model, while what is presented in the book is the first (and only) comprehensive critique of the logical empiricist and Popperian theories of science that attempts not only to show that both are dependent on the Deductive Model, but that both are fundamentally mistaken for that very reason.

The Transcendence of the Deductive Model

In the first five and a half chapters of the book, on the basis of what empiricists and Popperians say and do themselves, I reconstruct their philosophies of science in terms of the Deductive Model. In fact I do more than merely reconstruct their views; I show that their very thinking presupposes the model. This is so in the case of all their central concepts, including explanation and prediction, and the empiricists' notions of verifiability, induction, confirmation, progressive theory change, theoretical terms and correspondence rules, as well as the Popperian notions of falsifiability, basic statement, background knowledge, corroboration, severity of tests, theory conflict, theory content, probability of a theory, testability, verisimilitude, and the whole of Lakatos' sophisticated methodological falsificationism. I then go on in Chapter 6 (and briefly in Chapter 12) to show in the case of Popper and Lakatos that a number of their claims are merely descriptive and not explanatory – either because in those claims they use concepts such as 'content,' which are *incoherent* on their philosophy of science,[7] or because they use concepts such as 'new idea' in a way that is *antithetical* to their philosophy of science,[8] or because they use concepts such as 'one theory correcting another' which have not been *explicated* in their

---

[6] Shrader (1985), p. 223.
[7] Cf. *this volume*, pp. 28–34, 42 and 45–46.
[8] Ibid., p. 46.

philosophy of science.[9] As I say in introducing this line of criticism at the end of Chapter 5, "it will be found that to the extent that Popper and Lakatos rely on the Deductive Model they do not avoid the criticisms of the present section, and that to the extent that they do not, they no longer explain science, but merely describe it." In this regard Blackmore also praises my acuteness in pointing out "the manner in which many of Popper's ideas are not consistent with his own reliance on the Deductive Model."[10] Note that an appreciation of my criticism here requires an understanding on the part of the reader of the *coherence* that must be displayed by a philosophical theory.

Gärdenfors maintains that, on the basis of my assumption that Popper and Lakatos are working within the Deductive Model, I criticise them in Chapter 6 for the fact that their conceptual constructions cannot be described within it. But, continues Gärdenfors, Popper and Lakatos never intended to limit themselves to the Deductive Model, and definitely not in the form in which I present it.[11]

But it should have been clear to Gärdenfors at least by the time he arrived at my criticism of Popper and Lakatos for their transcendence of the Deductive Model that I was not *assuming* them to be working within it, but taking them to be committed to it *on the basis of my previous analysis*. Furthermore, Gärdenfors' claim that Popper and Lakatos never *intended* to limit themselves to the Deductive Model is not only moot but irrelevant. And when he says, 'definitely not in the form in which I present it,' I must ask: in what other form then? The form in which I present the Deductive Model captures all of the central concepts in the Popperian philosophy of science. I cannot imagine any other form that can do this; and Gärdenfors' not indicating what that other form might be makes what he says here vacuous.

Gärdenfors continues:

What Dilworth misses here is that Popper, and to an even greater extent Lakatos, complement deductive logic with an analysis of the *methodological decisions* that researchers make, consciously or unconsciously, to decide what is to be accepted as the empirical basis of a theory. Lakatos distinguishes between 'dogmatic' falsificationists who presuppose that there exists an objectively given empirical basis that can be proved from observable facts,

---

[9] Ibid., p. 129; quote following, p. 40.
[10] Blackmore (1984).
[11] Gärdenfors (1982), p. 45; quote following, ibid.

and 'methodological' falsificationists who acquire the 'empirical basis' of a theory through a series of methodological decisions .... Dilworth writes as though Popper and Lakatos were dogmatic falsificationists, which they of course are not.

But I certainly do not *miss* that Popper and Lakatos complement deductive logic with considerations regarding methodological decisions and so on. To deal with this issue is one of the main purposes of Chapter 6! And in fact on page 44 I even anticipate Gärdenfors' criticism, saying that it is naïve! Furthermore, not only am I *aware* of Lakatos' 'methodological falsificationism,' in Chapter 6 I first *quote* it, and then *analyse* it in terms of the Deductive Model.[12] And how can Gärdenfors say that Popper and Lakatos "of course" are not dogmatic falsificationists, when, for example, Kuhn has said regarding Popper: "Having barred conclusive disproof, he has provided no substitute for it, and the relation he does employ remains that of logical falsification. Though he is not a naïve falsificationist, Sir Karl may, I suggest, legitimately be treated as one." While I have not missed anything of what Gärdenfors mentions, he has himself missed (among other things) the whole of my criticism of Popper and Lakatos for transcending what is permitted on their own philosophies of science.

While Elzinga admits that my critique of Popper in Chapters 3 and 5 may be correct, he claims that it is only true of the very early Popper.[13] But this must be wrong, for, as I cite in the notes, the works of Popper's that I refer to were published between 1934 and 1975 – not his early work, but work produced during the *whole* of his career. And Popper has nowhere repudiated the views I take up.

What Elzinga considers interesting about Popper's and Lakatos' work is that they provide a *heuristic of research programme development*, which constitutes the core of a critical rationalism. Were one to recognise that this is what they do, he suggests, "one would also have to admit that Lakatos and even Popper to some extent have programmes that focus on scientific change, and as such afford a partially rational criticism of scientific progress." And as regards my

---

[12] *This volume*, pp. 41–45; quote following, Kuhn (1970b), p. 14.
[13] Elzinga (1981–1982), p. 267; quote following, ibid.

efforts to reconstruct Popper's philosophy of science in terms of the Deductive Model, I simply misrepresent it.

As regards Popper's and Lakatos' affording "a partially rational criticism of scientific progress," neither Popper nor Lakatos are critics of scientific progress, and I cannot imagine where Elzinga gets the idea that they are. Furthermore, that they should afford a *partially* rational criticism suggests that they also provide partially *irrational* criticism, though this of course is not what Elzinga means. But what he does mean is not clear. And as regards my misrepresenting Popper's philosophy of science, I am very careful to give references to Popper's text when I reconstruct his view in terms of the Deductive Model. Where, exactly, do I misrepresent him?

Elzinga goes on to suggest that I admit that the contradiction between the 'reconstructed' Lakatos and the real Lakatos becomes too obvious when I say that he is "fairly successful in making rational evaluations of scientific theory-change and progress" and then add that to the extent that he is successful, he departs from the covering-law model. Elzinga feels however that my claim that even Lakatos bases his reconstructions on the covering-law model is not convincingly argued.[14]

But my suggesting that Lakatos illegitimately transcends the Deductive Model is not an *admission* on my part, but both an attempted explanation and a criticism. Elzinga should have said that I attempt to *explain* particular of Lakatos' pronouncements and *criticise* him on the basis of this explanation, and having said this then argue that my attempted explanation is unsuccessful and/or that my criticism is unwarranted. Like Gärdenfors, Elzinga neither presents my position nor comes to grips with it. And I nowhere say that Lakatos "is fairly successful in making rational evaluations of scientific theory-change and progress," as Elzinga claims. Furthermore it is of course not here a question concerning the covering-law model, but the Deductive Model; but I nowhere suggest that Lakatos bases his own reconstructions (which?) on either model.

Ironically enough, the general thrust of what Elzinga is saying here, according to which Popper and Lakatos are attempting to provide 'partially rational heuristics' rather than completely rational

---

14 Ibid.

theories, would support an anti-Popperian position that is even stronger than the one I am taking. Not only would *certain* things that Popper and Lakatos say then be merely descriptive and not potentially explanatory, but *all* of what they say would be merely descriptive (or prescriptive), and *nothing* would be potentially explanatory.

Similarly to Gärdenfors and Elzinga, Shrader suggests that:

> Dilworth admits the difficulty of fitting Lakatos' views into the Deductive Model, but criticises the part which won't fit as ad hoc. He refuses Lakatos proper rights to any views beyond that entailed by the Deductive Model, despite Lakatos' claim to be a '*sophisticated* falsificationist.' Lakatos' sophistications are disallowed even when they are "almost invariably quite correct."[15]

It would perhaps not be fair at this stage to suggest that my critics have preconceived ideas regarding the questions at issue, but that is certainly the impression one gets. Again, I do not *admit a difficulty for my view* as regards Lakatos' transcendence of the Deductive Model, but *point out a difficulty for Lakatos' view*. And I do not deny Lakatos the right to views beyond that entailed by the Deductive Model; nor do I disallow some of what he says. Rather, I claim that some of what he says is only descriptive and not potentially explanatory: relative to his philosophy of science it is ad hoc.[16] So *is* it ad hoc or isn't it? Shrader doesn't even try to show that it is not. Nor does he – or Gärdenfors or Elzinga – mention that Lakatos fails to provide a conception of scientific progress even if he is thought *not* to presuppose the Deductive Model. And, just like Gärdenfors, Shrader manages to miss the fact that I *treat* of Lakatos' sophisticated methodological falsificationism. How is this possible?

Possibly apart from Ross Phillips,[17] my critics are quite unable to appreciate the difference between providing a general philosophical *description* or *analysis* and providing a philosophical *theory*. In the present book there are instances of both. For those who cannot see the difference, I point out that general philosophical analyses are

---

[15] Shrader (1985), pp. 224–225.

[16] I similarly suggest that certain views expressed in the later writings of Carnap, Nagel and Hempel are ad hoc: cf. *this volume*, p. 99n.; and I criticise Sneed for some of the things he says for the same reason: ibid., pp. 115–116&n.

[17] Who notes that "the purpose of the book is to launch a new theory." Phillips (1983), p. 214.

performed in Appendices I, IV and VII; and philosophical theories are presented in Chapters 1 to 5 (logical empiricism and Popperianism), Chapters 8 and 9 – with applications in Chapters 10, 12, and Appendices III, V and VI (the Perspectivist conception), Chapter 11 (the set-theoretic conception) and Appendix VIII (a theory of identity and reference). This inability on the part of my critics is perhaps supported by the logico-linguistic tradition, for in it, just as no distinction can be made between what is and is not ad hoc, so no distinction can be made between explanation and mere general description.

2. THE GESTALT MODEL

Phillips appears to have understood the nature and purpose of the Gestalt Model where he says that it and the Deductive Model are bases in two different senses: the Gestalt Model attained the status of constituting the basis of the Perspectivist conception due to its heuristic value, while "falsificationism and logical empiricism are plainly related to their basis in a much more intimate way." And his understanding continues where he goes on to note: "The presentation of the Gestalt Model simply consists in suggesting a number of features of gestalt-switch perceptual phenomena, analogues of which we will be encouraged to find in cases of theory change"[18] – though here he should have added: "within the context of the Perspectivist conception."

Phillips promptly loses this glimmer of understanding, however, and goes on to say that there is an illusion of explanation in my presentation of the Gestalt Model. That there might only be an *illusion* of explanation is hardly surprising, since the model is not intended to explain anything, but rather, as Phillips has himself said, to function simply as an analogy to aid understanding (of the Perspectivist conception). Phillips goes far with this misconception:

It would seem that our inability to see the duck-rabbit simultaneously as both a duck and a rabbit is to be explained by the joint occupancy by 'duck' and 'rabbit' of the same predicate category together with the explication of 'seeing as' as the application of a concept. ... It is ... not obvious that 'picture-duck' and 'picture-rabbit' are incompatible concepts in the required sense.

---

[18] Ibid.

... In any case Dilworth seems undecided about just what status our inability to see the figure as a duck and a rabbit is to have. This talk of predicate categories suggests that it is a matter of necessity, but in other places he speaks as if it is just a matter of psychological fact.[19]

But what is the "required" sense in which 'picture-duck' and 'picture-rabbit' are to be incompatible? All that is *required* is that the Gestalt Model provide an intuitive empirical understanding of the central concepts in the Perspectivist conception. What the "status" is of the fact that we cannot see the duck-rabbit as both a duck and a rabbit simultaneously is not a question that need be dealt with in the presentation of the model. As I say in the text: "Note ... that a treatment of the question as to *why* the duck-rabbit has the particular features it does is perhaps best handled by a psychologist; at any rate, such a treatment lies beyond the scope of the present study." How has Phillips missed my saying this? That he believes the Gestalt Model to have an explanatory function suggests that he has a fundamental misunderstanding of what is going on in the book.

Roy Bhaskar admits that, from the Gestalt Model, I am able to elaborate criteria for choosing between conceptual schemes applicable to a common domain, such criteria being simplicity, scope and accuracy.

There is, however, an important disanalogy that Dilworth fails to identify. In the simple gestalt switch case, the percipient is able to view the figure indifferently under its various aspects, and to switch more or less freely between them. But this is of course precisely what Kuhn and Feyerabend deny is always or even typically possible in scientific revolutions. Dilworth's Gestalt Model thus begs what they would regard as the main point at issue.[20]

Bhaskar is here considering two features of the Gestalt Model, the first apparently being a person's ability to see the figure under its different aspects without judging one of the aspects to be superior to the other, and the second being a person's ability to switch more or less freely between them. But I do not fail to identify the point Bhaskar raises, as he claims. Not only do I identify it, I claim it not to be a *dis*analogy but an *analogy* to the scientific case:

---

[19] Ibid; next quote, *this volume*, p. 58, n. 10.
[20] Bhaskar (1983), p. 259.

It may be thought though that where in the Gestalt Model one can easily shift back and forth between perspectives, this is not so in the case of scientific theories. But, while it may take some time before a scientist grasps or *understands* a new theory, the actual shift from not understanding to understanding, a step which may occur more than once before the new theory or perspective is internalised, is clearly similar to the dawning of an aspect which occurs in the Gestalt Model. ... And, once the scientist has understood both theories, though he might believe one of them to be clearly superior to the other, he can nevertheless easily shift from working within one to working within the other, just as a person can easily shift from one aspect of the Gestalt Model to the other.[21]

Why does Bhaskar, rather than direct his comments to what I say in this regard, claim that I fail to identify the point?

To the above it may be added that neither Kuhn nor Feyerabend denies that a person can work within a superseded theory without judging it, nor that one can switch more or less freely between pre- and post-revolution theories. Furthermore, neither considers such a question to be "the main point at issue," as Bhaskar's non sequitur suggests.

3. THE GESTALT MODEL VS. THE PERSPECTIVIST CONCEPTION

Regarding the relation between the Gestalt Model and the Perspectivist conception of science, I say in Chapter 9:

So here scientific theories are not taken to be statements having a truth-value, but are instead likened to individual empirical concepts or predicates which are intended to apply to certain phenomena – predicates such as 'red' or 'round,' or, with reference to the Gestalt Model, concepts such as 'picture-duck' and 'picture-rabbit.' Thus theories are here conceived of as being intended to apply to certain states of affairs, and to be such that they may be judged to be more or less successful in their application.

And both in the book's introduction and at the beginning of Chapter 8 I clarify that the Gestalt Model/gestalt-switch figure is being advanced in this work especially to serve as the (intuitive) basis of

---

[21] *This volume*, p. 70; as regards both of the features Bhaskar takes up, see also p. 211&n.

the Perspectivist conception.[22] And in Chapter 9, under the heading 'The Perspectivist Conception *as based on* the Gestalt Model,' I point out that:

The basis for the alternative conception to be presented below is, in this study, being taken as the Gestalt Model. It may be noted however that a different starting point might have been taken, such as, for example, a comparison of colour concepts, or a direct treatment of theories themselves. It is because so many features of gestalt-switch phenomena may be seen to have counterparts in the comparison of rival scientific theories that the heuristic value of these phenomena has resulted in their here attaining the status of constituting the basis of the present view. The following presentation will thus parallel the presentation of the Gestalt Model in the previous chapter, but certain points to be made will occasion the use of other sorts of examples and analogies.[23]

And I further point out in a footnote that· "The present conception as a matter of fact originated independently of the Gestalt Model." (My reason for introducing the Gestalt Model into the presentation of the Perspectivist conception was the difficulty earlier commentators had had understanding it;[24] and I thought that by linking its notions to empirical reality through a gestalt-switch figure this problem might be overcome.) As it *turned out*, the gestalt-switch phenomenon provides a very good empirical analogy to the Perspectivist conception – and thus a very good way of introducing it – in that it is easier to understand while still containing all the essentials. But the Perspectivist conception is in no way dependent on it.

Despite my saying all the above, however, Shrader nevertheless complains that the exact (*sic*) relationship between the Perspectivist conception and the Gestalt Model is unclear.[25] Even worse, Elzinga believes that the book's "central idea is that of the gestalt switch."[26] Meanwhile Gärdenfors makes the mistake of thinking that Chapter 10 of the book is to constitute an illustration of a gestalt switch in

---

[22] Ibid., pp. 2 and 56; in other terms, the Gestalt Model is being presented as the *source-analogue* of the Perspectivist conception: in this regard see e.g. ibid., pp. 187–188.
[23] Ibid., pp. 68–69; next quote, p. 69n.
[24] As it was presented in Dilworth (1976).
[25] Shrader (1985), p. 222.
[26] Elzinga (1981–1982), p. 266; next quote, Bhaskar (1983), p. 258.

science! And Bhaskar, for his part, quite mistakenly claims that my "central contention is that the different aspects of a gestalt switch figure ... provide a good model for the relation between competing or successive theories." This can only mean that these reviewers have not read the book carefully enough.

4. THE PERSPECTIVIST CONCEPTION

Bhaskar, in his critique, says that I am only able to avoid a particular problem – which is apparently that of comparing the relative adequacy of theories – by tacit recourse to a variant of standard empiricism, as I employ the notion of (the results of) "uninterpreted operations" as a neutral theory-free measure.[27]

As regards the problem of accounting for scientists' determination of the relative adequacy of competing theories, i.e. the problem of accounting for scientific progress, the response on the part of the Perspectivist conception is that it can be determined on the basis of such criteria as accuracy, scope and simplicity. This presupposes a common empirical basis, the reasonableness of the existence of which is established first by showing how different gestalt figures have a common empirical basis that can be described independently of them, and then by suggesting that on the Perspectivist conception more generally there also exists such a basis – in the case of modern science one consisting in the results of *measurement*. But nowhere do I use the expression "uninterpreted operations," as Bhaskar quotes me as doing. Nor, I would suggest, is this view of the empirical aspect of science a variant of standard empiricism, as Bhaskar claims. On the standard empiricist view, modern science is based not on measurement but on *sense-data*.[28] But, as I say: "On the present conception, however, science is not seen to rest ultimately on that which can be directly experienced – subjective sense-data, but on the results of operations involving the employment of certain instruments – results which can be intersubjectively shared." The conception of the empirical aspect of science on the Perspectivist view is more similar

---

[27] Ibid., p. 259.
[28] With regard to the difference, see *this volume*, pp. 102–103, 183–185 and Appendix VII; and Dilworth (2007), pp. 85–92. Quote following, *this volume*, p. 103.

to Bridgman's operationalism, as I allow where I say that my view nevertheless differs from his mainly in the stress he places on the meanings of scientific terms.[29] In any case, whether or not Perspectivism is similar to some other view in this regard is irrelevant as regards the viability of the Perspectivist conception itself. What is relevant is whether what I am saying is part of a coherent picture which constitutes the Perspectivist conception (which I claim that it is), and whether it is epistemologically problematic (which, as will be discussed directly below, it has so far not been shown to be).

After likening the Perspectivist conception of the empirical basis to that of the empiricists, Bhaskar goes on to say that, "as there can be no *guarantee* that common operations will *always* be agreed upon, and as such operations need to be brought under (putatively incommensurable) descriptions before they can ground choice, the Kuhn-Feyerabend problem remains." But what sort of guarantee does Bhaskar want? In science agreement on common operations in the case of competing theories is seldom a problem. And as to whether the holding of incommensurable theories implies describing their common operations differently, as pointed out by H. R. Post, it does not.[30]

Bhaskar continues that, by suggesting the empirical aspect of science to rest on intersubjective results of measuring operations, I dehistoricise and desocialise science "in the manner of pre-Kuhnian philosophy of science." But I do not "dehistoricise" science.[31] Keeping in mind that I am doing philosophy of science and not history of science, my illustrations of the Perspectivist view are nevertheless all based on historical examples. And as regards my "desocialising" science, just because the fashion in the early 1980s was to mix sociology of science in with philosophy of science, philosophy of science free of sociological aspects still existed, as it still exists today. In any case, I state in the text that the book "is not directly concerned with the psychological or sociological aspects of science, but is to fall

---

[29] Ibid.; quote following, Bhaskar (1983), p. 259, emphasis added.

[30] *This volume*, pp. 139–140; quote following, Bhaskar (1983), p. 259.

[31] Rossi in fact sees the book as constituting "a precise clarification of what one means by 'epistemology of history'" (1982, p. 186).

wholly within the realm of what is normally considered to be the philosophy of science."[32]

What I am doing is advancing a *theory* or *philosophy* of science. This theory is *free* of sociological factors while *taking account of* historical ones. I should suggest that this *is* philosophy of science, not pre- nor post-Kuhnian philosophy of science. That my philosophy of science suggests the existence of an actual empirical (mensural) foundation for science is part of its nature, and is in keeping with both common sense and what scientists themselves assume.

Gärdenfors, speaking of incommensurable theories, says that part of what I mean by two theories' in the same area being *incomparable* is that with their help one *sees* empirical phenomena from different perspectives.[33] But not only would I never express myself in the terms Gärdenfors uses here, but I nowhere say what Gärdenfors in his own way says that I say. First, I never speak of two theories being "in the same area" (whatever that may mean), but rather as having or not having the same intended domain. Second, I never speak of incommensurable theories as being incomparable, but quite the reverse, where I say "the move from one scientific theory to its successor ... may be said to involve a 'shift of conceptual perspec- tive' – a shift which does not preclude a meaningful comparison of the perspectives concerned."[34] And I cite Kuhn in this context, where he says, "Most readers of my text have supposed that when I spoke of theories as incommensurable, I meant that they could not be compared. But 'incommensurability' is a term borrowed from mathe- matics, and it there has no such implication;" and at that place I give a page reference to Larry Laudan where he evinces such a misunder- standing. Despite this, Gärdenfors attempts to express my view in terms of incomparability.

And third, and perhaps most important, not only do I not speak of *seeing* phenomena differently from different perspectives, but the very first section of the chapter on the Gestalt Model is devoted to this question, saying that the model is to function as an analogy on a conceptual rather than a perceptual level. And in the second section of that chapter I underline this by saying: "In contradistinction to the

---

[32] *This volume*, p. 67.
[33] Gärdenfors (1982), p. 45.
[34] *This volume*, p. 69; next quote, p. 69, n. 5.

use of [a gestalt-switch] figure in the context of discussing certain perceptual phenomena, emphasis will here be placed on particular *conceptual* features that it exhibits."[35]

Phillips makes the same sort of mistake where he says that the success of my treatment of a particular issue (viz., the problem of meaning variance) will depend on how well the notion of aspectual incompatibility, developed in the Gestalt Model, can be applied to scientific theories. And he maintains that the fact that the gestalt-switch case involves fairly elementary visualisation while developed scientific theories typically do not suggests that it cannot be so applied.[36] But the *perceptual* aspect of the Gestalt Model is part of its *negative analogy*[37] vis-à-vis the Perspectivist conception. I *abstract* from this perceptual aspect; and the determination of whether the notion of aspectual (rather, perspectival) incompatibility can be applied to developed scientific theories requires consideration of my actual attempts in the text to apply it to them.

Perspectival Incompatibility

Perspectival incompatibility is a conceptual clash between different perspectives that have the same intended domain; and, in being dependent on the nature of the central concepts in the relevant perspectives, it may occur even when those perspectives suggest the same results. With regard to this notion, *Stellan Welin* says:

Dilworth believes that theory conflict most often means that we have two theories for the same area and that the theories are equally compatible/ incompatible with observations. ... Is it really the case that there exist examples of different theories with exactly the same predictions? Doesn't Einstein's special relativity theory make essentially different predictions than Newton's mechanics? ... I'm inclined to agree with Popper that these new and different predictions (that show themselves to be correct) constitute a great step forward. More has happened than that we have only changed perspective.[38]

---

[35] Ibid., p. 57.
[36] Phillips (1983), p. 215.
[37] Cf. *this volume*, p. 144.
[38] Welin (1982), p. 24.

To begin with, Welin is unable to distinguish philosophical theorising from the making of factual claims. I am saying that *on the Perspectivist conception* theory conflict is not dependent on the suggesting of different results. This is part of the presentation of my theory, not a descriptive claim about science. Thus whether it is "really the case that there exist examples of different theories with exactly the same predictions" is not the point. Though I in fact give examples of such theories[39] (which Welin ignores), what is key is that on the Perspectivist view the *essence* of theory conflict does not lie in the provision of differing results. And of course I do not "most often mean" that theories "for the same area" (?) are equally compatible with (the results of) observations, neither as a factual claim about science nor as what should be the case on the Perspectivist conception. That Welin should think I mean this suggests an extremely superficial reading of my text. And when he says that new and different predictions which are correct constitute a step forward – which is in keeping with the Perspectivist view – he shows not only that he has not understood the Perspectivist conception, but that he has been unable to distinguish the issue of theory conflict from that of relative acceptability.

Meaning Variance

The problem of 'meaning variance,' i.e. that (type) terms common to two competing theories should differ in meaning in each of the theories, is a problem for philosophies of science based on the Deductive Model, which, while they see the epistemologically relevant relations between competing theories to be linguistic, can only handle syntactic and not semantic relations.

Welin's superficial reading is also manifest in his comments on my treatment of this subject. In this regard he says that I believe that the fact that we can *measure* impulse in the same way in (*sic*) both Newtonian mechanics and Einsteinian relativity to constitute a *linguistic discovery*; and more generally he suggests my position to be

---

[39] *This volume*, pp. 74–75 and 96, n. 9. For further examples, added after the first edition, see pp. 75n. and 200–201.

such that terms always change their meanings in the case of theory succession.[40] (Which terms? *All* terms?)

First of all, I nowhere suggest that theory comparison involves linguistic discovery. And I do not assume that theory change necessarily involves meaning change. As I say in this latter regard, "while the present view *allows* for the meaning variance of individual terms ... it is the theories as wholes that are here seen to conflict[. S]hift of conceptual perspective ... may or may not be accompanied by a meaning change of individual terms."

Ignoring or missing this, Welin goes on to defend the Deductive Model in its guise as the D-N model, suggesting that if terms change in meaning then such a change should occur in the case of all terms of the same type in a particular explanation. And he concludes, saying, "I don't understand the way in which the maintenance of the Deductive Model should mean that these problems of meaning change and so on should be difficult to discover."[41]

But the issue does not concern the 'discovery of problems of meaning change,' but what meaning change implies as regards the applicability of the Deductive Model. And even apart from the fact that Welin's example doesn't concern relations between theories (conceived as statements), it doesn't avoid the problem, for, in being purely syntactical, the Deductive Model cannot *in any employment* ensure that semantic change has not taken place. Thus it cannot ensure that all terms of the same type in any one explanation all have the same meaning – and if they do not, the model's formal relations are not merely irrelevant but misleading.

Phillips devotes much attention to my treatment of meaning variance, suggesting that I *trivialise* the problem. But the problem of meaning variance is one for philosophies of science based on the Deductive Model, and for them it is not trivial. For the Perspectivist conception meaning variance is simply not a problem, trivial or otherwise.

Phillips bases his claim of trivialisation on my saying with regard to meaning variance:

---

[40] Welin (1982), p. 23; quote following, *this volume*, p. 77, emphasis added.
[41] Welin (1982), p. 23.

If we approach this problem first on the level of basic epistemology, and take the meaning of a (descriptive) term to be the (empirical) concept it is being used to express, earlier considerations suggest that in certain cases conflict can occur precisely *because* there is a change in the meaning of such a term. If one person says that something is red, and is using the term "red" as it is normally used, while another person says that the same thing is red, but in saying so actually means 'blue' by the term "red," then we have an instance of the sort of conflict earlier called 'contrariety.'[42]

And I then go on to say that, in any case, "on the present conception theory conflict is not viewed as being based on the inter-theoretical relations among the meanings of individual terms, but is seen rather as resulting from the attempt to apply the whole of certain distinct theories to the same state of affairs."

With regard to the above example, Phillips says: "This is just a misunderstanding of the problem, for it does not arise where we may be presumed to have independent access to the meanings of terms, but rather it is a problem for those who want to say that theoretical terms get their meanings from the total theories of which they are parts."[43]

I must say I find Phillips' remarks here very difficult to understand. My example is an illustration of how, in a certain context, meaning change can actually *account* for conflict (a linguistic example which has important similarities to perspectival incompatibility). Meaning variance is not a problem for the Perspectivist conception, and my example is certainly not intended to solve the problem for those who presuppose the Deductive Model, whether they take theoretical terms to obtain their meanings from the theories in which they occur or not. In any case, why should the question of how one acquires knowledge of the different meanings of one and the same term make any difference (if this is what Phillips means by "independent access")? The meanings can still be compared.

Further pursuing this issue – which is not central to the Perspectivist conception – Phillips claims that I maintain that "the meanings of terms in experimental laws are linked to mensural operations." But I do not maintain this at all. I maintain the opposite: "The major difference between [Bridgman's] view and the present one is the stress

---

[42] *This volume*, p. 76; next quote, ibid., p. 77.
[43] Phillips (1983), p. 215.

he places on the *meanings* of scientific terms, and with it his suggestion that their meanings are ultimately determined by the operations
performed in their application."[44] Why does Phillips say that I maintain something when I expressly deny it?

Relative Acceptability

In the presentation of the Perspectivist conception three sorts of
criteria are advanced as potentially playing a role in determining the
relative adequacy of competing perspectives: accuracy, scope and
simplicity. As I say in the text, scientific theories are "to be thought
of as comparable with respect to these factors, and on the basis of
such a comparison may be *judged* as to their relative acceptability;"[45]
and that these factors "may play a role in the possible *argumentation*
concerning the relative acceptability of competing theories." (As
expressed by P. A. Rossi with regard to these criteria, on the Perspectivist conception "scientific theories are ... structures having a
greater or lesser degree of applicability depending on the results of
the particular 'mixture'."[46]) And I say:

> Of course there is nothing to prevent more complicated situations from
> arising in which, for example, one theory has a wider scope than the other,
> but the other is the more accurate in application to the domain to which both
> more or less apply. And it is clear that in such a case the debate over which
> of the two theories is all in all the more acceptable cannot be so easily re-
> solved; and both theories might be retained, each to be used in those par-
> ticular cases where the sort of superiority it evinces can come to the fore.[47]

Writing as though I haven't said all this, however, and ignoring
the fact that I here speak of *criteria*, Shrader asks:

> What happens if one theory (*A*) makes more accurate predictions than
> another (*B*) as regards one sort of phenomenon, but *B* makes more accurate
> predictions than *A* with respect to another? Again, what if *A* has greater
> scope than *B*, but *B* is more accurate than *A*? Dilworth offers precious little

---

[44] *This volume*, p. 103.
[45] Ibid., p. 85; next quote, p. 86; emphasis added in both cases.
[46] Rossi (1982), p. 188.
[47] *This volume*, p. 86.

guidance for such cases, beyond suggesting that perhaps we should retain both A and B.[48]

Shrader apparently expects me to provide an *algorithm* for determining which of any two competing theories is superior to the other. But, as I have pointed out, theory choice is, on the Perspectivist conception, precisely a matter of *choice*, more particularly of *judgement*, and in certain cases that judgement may be difficult or impossible to make – just as it is in actual science. An algorithm for theory choice is what philosophies of science based on the Deductive Model (unsuccessfully) attempt to provide. The fact that Shrader cannot see that I am doing something else suggests that he is unable to step back from the deductivist way of thinking in considering my alternative, and that this has prevented him from understanding that alternative.

5. APPLICATIONS

As intimated at the beginning of Chapter 10, my purpose in that chapter is to further explicate the Perspectivist conception in the context of the theory of gases. Thus, apart from showing how *all* of its central concepts are to apply to the Bernoulli/van der Waals case, I there state that on the Perspectivist view scientific theories are to be conceived as *applied models*, and I introduce into the Perspectivist conception such notions as those of *reduction, limiting case* and *idealisation*. And, in Chapter 12, I apply the Perspectivist conception to the views of Newton, Kepler and Galileo (and, in later editions of the book, to further examples). Here I might also point out that these applications to actual science are more detailed than those of any other theory attempting to account for scientific progress.

Shrader advises his readers that, because these examples are logical reconstructions that fail to develop or expand the Perspectivist conception in a philosophically interesting manner, they can be omitted or skimmed.[49] But first, they are not *logical* reconstructions; and second, if Shrader does not find them to be philosophically interesting he should say why. Since he elsewhere says that the first six chapters of the book can also be omitted, according to his advice a

---

[48] Shrader (1985), p. 224.
[49] Ibid.

full understanding of the text should be obtainable by only reading
Chapters 7 to 9! Not only is this absurd advice to give his readers,
but it is highly ironic, since Shrader himself has been unable to
understand the book despite his ostensibly reading the whole of it.

Elzinga, for his part, claims that my training as a logician (?) has led
me into anecdotal treatments of the history of science, and that my case
studies add nothing to what Kuhn and Feyerabend have already said.

It seems funny to start with a big question like gestalt switches in paradigm
changes and then end with a little question concerning the differences between Boyle's and van der Waal's laws. From the point of view of the
working scientist there is no perspectival shift involved here, only different
approximations. If one wants to do a case study to illuminate the shift of
theoretical and conceptual-framework perspectives in a line of progressive
growth [*sic*] in science, the transition from Newton's to Einstein's theory of
gravitation is a much more appropriate subject.[50]

Both of Elzinga's claims, that my case studies are anecdotal and
that they add nothing to what Kuhn and Feyerabend have already
said, are also absurd. (That I provide a *theory* of scientific progress,
while neither Kuhn nor Feyerabend do, is completely missed.) I
know of no anecdotal passages in the whole of the book; and Elzinga
does not point out any. Nor does he indicate one thing that I say that
has already been said by Kuhn or Feyerabend. And I do not understand how it should seem "funny" to him that I start with the "big"
notion of a gestalt switch and then end (?) by applying the Perspectivist conception to the theory of gases (*not* to Boyle's and van der
Waals' laws, as he suggests). Should I not have started by introducing my theory and ended by applying it?

Elzinga is not alone however in thinking that the Perspectivist
conception does not apply well to the example from gas theory. As
mentioned, *Gärdenfors* believes the example of Bernouilli's and van
der Waals' gas models to be a poor illustration of a gestalt switch in
science, Bernouilli's model being (as I myself say in terms of
"reduction" and "limiting case,"[51] mentioned above) a special case of
van der Waals'.[52] And, similarly to Elzinga, he suggests that I could

---

[50] Elzinga (1981–1982), p. 268.

[51] *This volume*, pp. 95–96.

[52] Gärdenfors (1982), p. 46.

have illustrated perspectival change much better by taking up some of the scientific revolutions that Kuhn has studied. And *Welin*, for his part, has a great deal of difficulty in seeing the analogy with the duck-rabbit in my comparison of Newton and Kepler.[53]

What these commentators miss, however, is that the Perspectivist conception is, as I mentioned earlier, conceptually independent of the Gestalt Model; and furthermore, unlike Kuhn's and Feyerabend's intentions regarding the incommensurability thesis, as I say at the end of Chapter 7, my intention is "to provide a general characterisation of theory change which is applicable to *any* case of competing scientific theories."[54] And even more important, what they also miss – as parenthetically noted above in the case of Elzinga – is the fact that what I am doing is providing a theory of science and not simply a philosophical commentary regarding its nature (a point which can account for many of their misunderstandings). My intention in Chapter 10 being *both* to develop *and* to apply the Perspectivist conception, I do this in the context of an example which at first might be thought not to be covered by it, and which is both easy to understand and at the same time sufficiently sophisticated to allow for development. If I had followed the advice of Elzinga and Gärdenfors I would immediately have had to meet the criticism that, yes, the Perspectivist view applies to radical theory change, but I haven't shown it to apply to more commonplace theory change. The Perspectivist conception is an abstract conceptual scheme the applicability of which to a particular example is to be determined by looking at its various concepts and seeing whether they fit the example, not by having some sort of gut reaction to the effect that 'this doesn't seem like a gestalt switch to me.'[55]

---

[53] Welin (1982), p. 28; Welin apparently thinks that I consider Kepler's laws to constitute a conceptual perspective, which I expressly do not: see *this volume*, p. 125.

[54] Ibid., p. 54; emphasis added.

[55] Reviewers are not unanimous, however, in criticising my examples. Cf. Lauener: "I must not fail to point to the particularly interesting illustrations [of Dilworth's conception] in Chapters 10 and 12" (1983, p. 69); and Rossi: "Such examples are without doubt very well constructed, and in accord with Dilworth's vision of the nature of the relation between successive scientific theories. The provision of such examples ... gives even more support to the applicability of Dilworth's [vision], and recommends it to a more attentive discussion than this review might offer" (1982, p. 188).

6. CONCLUSION

Ironically enough, what I am offering is a view of science completely in keeping with most of my critics' conceptions of it,[56] but one which, unlike those of the empiricists, Popperians and structuralists, I suggest is both *coherent* and can *account* for how science develops. But one gets the impression that for some reason my critics are *affronted* by the fact that such an alternative should be advanced. Thus, to say the least, they hardly read the book with an open mind; in fact they hardly read it at all. What they do rather is jump at the first thing they don't understand or have missed, take it to be a failing on my part, and criticise me on this misreading.

Not only have my critics failed in their criticisms of this work, but they have also failed to reveal to their readers what the book's message is. Though it is perhaps to expect too much that they do so, since they seem not to have understood this message themselves, they might still have made the attempt; and, had they read the book sufficiently carefully to see that it presents a complete theory of science, they might have passed this fact on to their readers. Furthermore, they might also have taken up some of the questions of central concern to the book, such as whether it has succeeded in meeting Kuhn's and Feyerabend's claims while avoiding relativism.

An author should be able to expect that, whatever his reviewers' philosophical leanings, they take care and attempt to maintain some semblance of objectivity, especially when the topic they are dealing with is as important as that treated in this book. But not only the writers critical of the first edition of this work, but similarly the reviewers of later editions,[57] have taken very little care in reading the book, and have maintained very little objectivity. As a result, virtually all criticisms of this work are based on a skewed and superficial reading. More important, however, is that the nature of the critical treatment the book has received has stood in the way of its true value being revealed to those who have not themselves had the opportunity to read it.

---

[56] Cf. *this volume*, pp. 49 and 67.
[57] See Dilworth (2007), App. III.

PERSPECTIVISM AND SUBATOMIC PHYSICS

1. APPLIED NUCLEAR MODELS AS CONCEPTUAL PERSPECTIVES

The notion of conceptual perspective finds clear application to subatomic physics in the context of applied nuclear models. There exist a number of models of the nucleus, of greater or less sophistication. These models may be divided into two basic groupings: strong interaction models (SIM) and independent particle models (IPM).

Examples of SIM are the liquid drop model, in which the nucleus is conceived on analogy with a spherical drop of liquid which retains its shape due to forces among the particles constituting it, and the more sophisticated collective model, in which the internal forces cause a deformation in the spherical shape suggested by the liquid drop model. Examples of IPM are the Fermi gas model, which represents the nucleus on analogy with a gas consisting of relatively freely moving atoms, and its elaboration in the form of the shell model, which places greater restriction on the freedom of the motion of the nucleus' constituents. Finally, there is the unified nuclear model, which is an amalgamation of strong interaction and independent particle models, and is more complicated than any of the other models mentioned above.

Taking such models, in application, to constitute conceptual perspectives, we shall examine them in terms of the five notions central to the Perspectivist conception, thereby demonstrating one of its applications to subatomic physics.

Intention

In the case of applied nuclear models the intention or intended domain of each of the perspectives is the same, namely, atomic nuclei. But, as regards the Fermi gas model for example, we can see how the intended domain of application of essentially the same model might have been, and actually has been, quite different, namely, real gases.

Now, as a matter of fact, it is seldom if ever the intention of a scientist to apply a model to a state of affairs to which the model bears no correlation whatsoever. Nevertheless, we note that the intended domain of application of a perspective need not be completely explicable by that perspective. Continuing in our considerations of the application of the Fermi gas model for example, we note that this particular perspective, though it has atomic nuclei as its intended domain, is quite unable to account for certain of their basic features, such as specific properties of their excited states. Thus we see that that to which the model is applied, or, that which the perspective is on, is in an important sense independent of how well it actually explains the phenomena, while it is dependent upon the *intention* of the individual who has the perspective.

Categories

On the view taken here, and as is generally recognised, the basic empirical (mensural) categories or parameters which are of relevance to nuclear models are those of mass and charge. One way mass can be determined, given the knowledge of Avogadro's number, is by the x-ray method, which involves the use of an x-ray diffraction apparatus. Charge can be ascertained, for example, using an oil-drop apparatus such as that used in Millikan's experiment. Some further parameters of relevance concern the lifetime of particles, as may normally be measured with the aid of a counter; their spin, which may be indirectly determined, via the Zeeman effect, through the use of a spectroscope; and their magnetic moment, which may be determined with the help of an omegatron.

Simultaneity

If we consider strong interaction and independent particle models each in their extreme form, we obtain a clear idea of how such models cannot be applied simultaneously to the same intended domain by one and the same person. This is so because the one sort of model assumes nucleons to be strongly coupled to each other (SIM), while the other assumes their motion to be almost independent of one another (IPM); and these contrary characteristics cannot be attributed to the whole of some one entity at one and the same time.

The above does not mean, however, that certain compromises cannot be made between the two extremes, resulting in but one model resembling each of its predecessors to a limited degree, as is the case with the unified model. But what it does mean is that, so long as the applied models, or perspectives, are distinct, they cannot be treated as constituting but one system. Though we might, for example, determine an equation relating the intra-nuclear forces of the liquid drop model to the potential well required to preserve the nuclear shape on one of the independent particle models, such a determination would be quite meaningless when we come to consider the models each as applied systems.

Perspectival Incompatibility

The fact that one cannot simultaneously apply, for example, both a strong interaction and an independent particle model to the same domain is here to say that the two are perspectively incompatible. This incompatibility rests in these models' characterising nuclei in ways which are mutually exclusive, and it does not at all depend on the perspectives' suggesting different results of the same measuring operations. Thus even if it were determined that two such applied models were actually empirically equivalent, they would still be perspectively incompatible due to the different account each provides of why the data are as they are.

In the above way we see too that the sort of conflict which can arise between these models is not essentially one of contradiction. Neither model by itself actually asserts anything to be the case, let alone denies anything asserted by the other. Rather, each model, in application, provides a positive account of particular empirical phenomena, and it is as a whole that each conflicts with the other.

Acceptability

Not all nuclear models are equally acceptable, and, on the present view, their relative acceptability may be considered in terms of three basic factors.

Accuracy: Of the five models introduced at the beginning of this section, the unified nuclear model is the most accurate. Not only does it share all of the predictive successes of the other models, such as the shell model's ability to account for the particular stability of those nuclei which consist of certain 'magic numbers' of protons or neutrons, but it is also able to provide an explanation of why many nuclear quadrupole moments are larger than would be expected on the shell model.

Scope: An instance of difference of scope in the present context may be provided via a comparison of the empirical efficacy of the liquid drop and Fermi gas models vis-à-vis the shell model. Here we may say that the shell model, in that it is able to explain specific properties of excited nuclear states – while these properties are quite intractable on the other models – as well as explain those empirical features which the other models are also able to explain, has a greater scope than either of them.

Simplicity: In the present context the notion of simplicity perhaps applies best to the way in which the shell model is able to account for the magic numbers' being what they are. Taking into account the spin-orbit force, on the shell model one is able to explain not only some of the magic numbers, but all of them. In this way the shell model provides a coherent conception in which all of the relevant data can be explained without requiring the implementation of ad hoc hypotheses to handle particular cases. Though it may be argued that the introduction of the notion of spin-orbit force is itself ad hoc, the fact that this one notion allows the derivation of the correct magic number in *every* case may be cited as a justification for its being taken to represent an actually existing force. It is in this sort of way then that this perspective may be said to be simple (with respect to the magic numbers) in the sense intended here.

2. THE DISTINCTION BETWEEN LAWS AND THEORIES

The above example involving applied nuclear models is intended mainly to show how an analysis of theories in subatomic physics in terms of conceptual perspectives might proceed. In considering sub-atomic physics in these terms it is important to distinguish between theories or applied models on the one hand, and empirical laws on the other,[1] and to note that it is the former and not the latter which are here being considered as conceptual perspectives.

An example from particle physics which is relevant to this distinction concerns the current effort being made empirically to establish a great number of selection rules for determining the magnitudes which are conserved in the disintegration of particles. This investigation is proceeding solely on the empirical level as a result of the fact that there presently exists no theory to suggest what these rules are. Once the rules are empirically determined, however, the task will be to develop a theory capable of explaining them.

3. IMPLICATIONS FOR SUBATOMIC PHYSICS

One of the implications of the preceding considerations is that the discovery of a certain functional relationship between particular parameters, or the obtaining of results via the manipulation of certain previously established equations, does not suffice to provide one with an explanation of the phenomena under consideration. In order that such phenomena actually be explained, they must be understood in the context of some sort of physical model (perspective) which can itself be relatively easily grasped independently of them. This demand on the perspective means that, for example, the fact that a certain group of data has found a unifying *mathematical* representation is not sufficient to justify our saying that we understand these data; nor is the depiction of a model by this mathematical representation sufficient if the model runs counter to our intuitions, or for some other reason is unsound.

Considering the inherently problematic nature of the subject-matter of subatomic physics, a second implication of the foregoing, complementary to the first, concerns the extent to which experimental findings are presented in a way which presupposes current general theoretical presuppositions in the discipline. Since subatomic particles, by their very nature, cannot be seen or touched, all evidence we might acquire regarding them must be indirect. What this leads to however is, over time, the manufacturing of a particular conceptual

---

[1] For a deeper analysis of this distinction than that in Appendix I, see the next appendix.

picture of the nature of these particles, and the tendency to interpret and represent all empirical results directly in terms of this picture. This in turn blurs the distinction between the empirical and the theoretical, making even more difficult the task of understanding the nature of the subatomic world.

ON THE NATURE OF SCIENTIFIC LAWS
AND THEORIES

The question of the difference between scientific laws and theories has received remarkably little attention in the philosophy of science. N. R. Campbell is perhaps foremost amongst those who do empha-sise the difference, taking it up in his classic *Physics: The Elements* (1920). Ernest Nagel also considers the issue of importance and, with reference to Campbell, deals with it in his *The Structure of Science* (1961). The distinction has also been clearly recognised by Rom Harré in *The Principles of Scientific Thinking* (1970), though he has not concentrated on it as an issue in itself. Apart from these authors, however, few have devoted attention to the question of how scien-tific laws differ from theories, and what their respective essential natures might be.

The question would of course not be so important if it were found that laws and theories do not essentially differ – if theories were nothing but 'higher-level laws,' or concatenations of laws. In the present appendix, however, the attempt will be made to show that this is not the case, and that, rather, laws and theories constitute dis-tinct and fundamental categories of science, an appreciation of the nature of which is a prerequisite to a proper understanding of science itself. Thus, the viability of the distinction to be made here will have a direct bearing on such basic issues as those of confirmation and scientific progress, necessitating their reconsideration with an eye to whether the notions are being applied to laws or to theories.

Before embarking on the presentation of this conception, it may be pointed out that there should exist a third scientific category at least as basic as those of laws and theories, namely that of *principles*.[1] This appendix, however, is concerned mainly with laws and theories,

---

[1] For a depiction of modern science in which principles are brought to the fore, see Dilworth (2007).

and so principles will be treated only insofar as their employment bears directly on this topic.

1. THEORIES ARE NOT SIMPLY LAWS REFERRING
TO UNOBSERVABLES

Nagel, though he is concerned to distinguish laws from theories, nevertheless begins his analysis by taking theories simply to be laws of a particular sort, viz., to be 'theoretical laws.' Such laws are to be about unobservable matters, and take the form of 'assumptions' about e.g. the molecular constitution of water, the atomic constitution of chemical elements, or the genetic constitution of chromosomes.[2] These laws are to be contrasted with 'experimental laws,' the key to the distinction lying in the notion of observability.

While much of what Nagel says about the difference between experimental laws and theories is in keeping with the view to be presented here, his starting point of taking theories simply to be laws about unobservables is not. In fact, his own examples speak against this way of conceiving of theories. With regard to the molecular constitution of water, for example, we might ask: what sort of assumption about it does Nagel mean should constitute a theory? If it is one restricted to the *behaviour* of the molecules, it might be considered a (theoretical) law. But such an assumption would not suffice to constitute a theory, for the existence of the molecules must also be asserted, if only hypothetically. As will be evident from what follows, the conception of theories as laws referring to unobservables is simplistic and contributes to such problems as that of theoretical terms – problems which can be avoided on a more sophisticated conception where theories are seen also to perform an ontological function.[3]

---

[2] Nagel (1961), pp. 79–80.

[3] A similar view has been expressed by Ernan McMullin where, with reference to Hempel, he says: "The [theoretical] explanans is not primarily a set of laws but a postulated structure of entities, relations, processes." (1984), p. 206.

2. THEORIES PROVIDE EXPLANATIONS

One aspect of Nagel's (and Campbell's) treatment of theories which is adopted here however is the conceiving of them as having the function of explaining empirical laws. The notion of explanation has of course been paid a good deal of attention in the philosophy of science, and in recent years the classic covering-law model of explanation has come to receive more and more criticism.[4]

Two main lines of criticism of the model are, first, that explanation in science, in its paradigmatic form, does not consist of formal deduction; and second, that such explanation is not based on laws, but rather on theories. Both of these lines are supported in the present appendix, and part of its task will be to provide a positive conception of scientific explanation which is not formal and is based on theories rather than laws.

In attempting to provide such a conception, one may or may not ignore the 'pragmatic' aspects of explanation. Though the meaning of the term "pragmatic" has not been made completely clear by those who use it, it does appear that it should include the notion of intellectual satisfaction. But a philosophical analysis of explanation which ignores whether it provides intellectual satisfaction would seem rather sterile, if not quite beside the point; and what we find is that even those who advocate a 'nonpragmatic' approach attempt to justify their results by showing how they provide a framework in which purported explanations may be seen to provide intellectual satisfaction. As has been expressed by Peter Alexander, "the test of whether an explanation is satisfactory must depend somehow on what people accept as explaining. We must start from whatever is generally accepted, in wider or narrower contexts, as explanatory."[5] Suffice it to say that an analysis of explanation which succeeded in abstracting completely from its 'pragmatic' aspects – if such were possible – would be left without a link back to actual explanation, and could hardly be expected to provide insight in regard to it.

At the same time, however, when it comes to the provision of intellectual satisfaction human nature disposes us to be satisfied with

---

[4] Cf. e.g. Scriven (1962), p. 203; Harré (1970a), pp. 15–21; Woodward (1979), p. 61; McMullin (1984), p. 214; and *this volume*, pp. 6–7.
[5] Alexander (1963), p. 114.

what we receive, even when it does not quite fulfil our original anticipations – especially when what we receive is proffered by a child so prized as science. Thus we might distinguish between what would constitute an *ideal* scientific explanation, on the one hand, and those 'explanations' modern science is in fact capable of affording, on the other.

The key to what constitutes an ideal scientific explanation may be found in the writings of those great minds responsible for the Scientific Revolution. Following them, we may briefly state this ideal as consisting in explaining why the phenomenal world is as it is, and changes as it does, as a result of the motion of non-changing matter.[6] A further aspect of this ideal is that the matter's state of motion should be the result of its coming into contact with other matter. This state of affairs may be conceived of in either of two ways, one involving the notion of empty space, and the other not.

Now this ideal was frustrated at a very early stage due to its apparent inapplicability to gravitation. Nor is it clear how it should be applied in the case of social phenomena. And so amendments have been made. But what is of importance from the point of view of the present appendix is that an ideal having more or less the nature indicated constitutes a *paradigm* for scientific explanation.

Against this background we thus see that to speak of scientific explanation as involving what has simply been termed a 'reduction to the familiar' is not sufficiently specific, and may in fact be misleading. It is not anything whatever with which one happens to be familiar that can suggest the basis of a scientific explanation, but ought ideally to be something which involves the interaction of physical objects.[7] And, in addition to this, it should be kept in mind that that which is being explained may itself also be quite familiar, and that this alone does not make it any less in need of explanation.[8]

---

[6] Concerning this ideal, see e.g. Dilworth (2007), pp. 59–60, 200–203 and 262–263.

[7] Failure to recognise this is quite widespread. Hempel, for example, argues against the 'pragmatic' view of explanation as he conceives of it, saying that "what is familiar to one person may not be so to another" (1965, p. 430); Campbell too misses this point in his discussion of 'explanation by greater familiarity' (1920, p. 146). So too Hutten (1954), p. 289, and Graves (1971), p. 48.

[8] Nagel thus misrepresents the situation where he says that "explanations can be regarded as attempts at understanding the unfamiliar in terms of the familiar" (1961, p. 107).

The above then provides an indication of the way scientific theories are here seen to function. There exist certain states of affairs with which we are acquainted – perhaps on the basis of experimental investigation – but whose *modus operandi* we do not understand. A theory is framed which attempts to explain this behaviour in terms of hypothetical processes which, from a scientific point of view, we do consider ourselves to understand. Sometimes we are here obliged to stretch our understanding somewhat – but this is reflected in the extent to which we are willing to grant that the 'explanation' the theory provides is truly satisfying.

3. THEORETICAL EXPLANATIONS ARE CAUSAL

Here scientific theories are taken to explain phenomena or experimental laws by indicating the precise way in which they can be seen as being the result of particular causes, as this notion is understood in science. The paradigmatic scientific notion of causality, at least for physics, is that in which change is brought about by the action of forces.

Often, in the formulation of laws or the observation of phenomena, one is aware of the presence of a causal connection, but one is not aware of the means by which that connection is carried through. One feels sure, for example, that there is a causal link between being bitten by mosquitoes in tropical areas, and the subsequent appearance of malaria, or between an increase in the voltage of a circuit, and an increase in its current. But the presence of the conjunction of the phenomena themselves does not suffice to satisfy our understanding as regards why the one is associated with the other. And it is in such sorts of case that one theorises as to what the nature of the causal connection might be. As expressed by Rom Harré: "Far from the discovery of constant conjunctions being identical to the discovery of causal relations, they are taken as the signs of causal relations, which have to be filled out by describing the mechanism which accounts for the constant conjunction."[9]

As the above examples indicate, and as is implicit in the everyday use of the term, theorising involves a delving into the unknown.

---

[9] Harré (1970a), p. 110. Laws assumed to be causal but which are as yet lacking a theoretical explanation, Harré calls protolaws: cf. ibid., e.g. pp. 132 and 154.

Theoretical explanations are sought in those situations where the nature of the causal relation, as conceived in science, is not manifest; were it manifest, there would be nothing to explain. And the provision of a scientific explanation consists precisely in indicating what the nature of the causal relation is. In some cases, one can look to see whether the situation depicted by the theory actually obtains. But most often, at least at the time when the theory is first advanced, this is not practically possible, and so the correctness of the theory can only be determined by indirect means.

The line of thinking presented here is similar to that of Ernan McMullin, where he says: "Theory explains by suggesting what might bring about the explananda. It postulates entities, properties, processes, relations, themselves unobserved, that are held to be causally responsible for the empirical regularities to be explained."[10]

4. EMPIRICAL LAWS, NOT INDIVIDUAL PHENOMENA, RECEIVE EXPLANATIONS IN SCIENCE

As is implicit in McMullin's characterisation, in modern science, as a rule, it is regularities or laws that are explained, not individual phenomena. This state of affairs has been noted by a number of authors, among them William P. Alston, who has devoted a paper to the issue. Though Alston unfortunately misses the distinction between laws and theories, and with it the fact that explanations are paradigmatically provided by the latter, with regard to what it is that is explained he rightly says:

Physics tries to explain phenomena of locomotion, expansion, and electrical transmission; chemistry the rusting of metals and the souring of milk; psychology, differential rates of learning and depth perception. But the research physicist does not occupy himself with the explanation of the boiling of a particular kettle of water as such ...; he is concerned to explain the nomological fact that water (under certain conditions) boils at 212°F. ... The search for explanation that is essential to pure science does not begin until some law-like generalisations have already been established (or at least accepted), and then it is directed to the question of why those generalisations hold.[11]

---

[10] McMullin (1984), p. 210.
[11] Alston (1971), p. 18. In this regard see also e.g. Campbell (1920), p. 87; Nagel (1961), p. 103; Harré (1970a), p. 35; and McMullin (1984), p. 215.

One might actually take this line of reasoning one step further, and say that the facts provided by science are nothing other than the laws or regularities it has shown to obtain. And while individual phenolmena do play a role in science, it is the role of providing evidence regarding what the scientific facts are. In other words, we might say that it is only when an expression has been shown to be that of an empirical law that it is admitted to the factual part of science.

5. THEORIES PROVIDE UNDERSTANDING; LAWS PROVIDE KNOWLEDGE

As has been mentioned in passing above, one's aim in providing a theoretical explanation is to afford the person receiving the explanation with an *understanding* of why certain states of affairs with which he or she is acquainted are as they are. In the exact sciences these states of affairs are most often the holding of numerical laws. In attempting to explain these laws we go 'behind' their outward manifestation and postulate the existence of something responsible for them. We by no means *know* that the responsible agent is what we suggest it to be – and sometimes we may discover after a time that we were mistaken in our characterisation of it. But our aim in such a context is, in any case, not to provide *knowledge*, but a kind of hypothetical *understanding*. *If* the world is as we suggest it to be, *then* these facts, which are in principle accessible to everyone, follow as a matter of course.

The essence of this line of thinking has been captured by Ludwig Boltzmann, where he says:

In order to *understand* the phenomena which actually occur, we may draw conclusions from hypothetical assumptions, that is, from processes which, though possible on analogy with similar phenomena in other circumstances, cannot be observed and may not even be observable in the future, owing to their speed or small size or something similar.[12]

---

[12] Cited in Flamm (1983), p. 261, from a lecture given by Boltzmann in 1903; emphasis added. Basically the same distinction as that being drawn here has also been made by Peter Alexander; in his words: "The aim of explanation is to achieve understanding, to make things intelligible, whereas the aim of description is to say how things are." (1963), p. 138.

In broad outline, the picture of science being sketched here shows it as having two basic tasks: one, the determination of what the facts are, and the other, the explanation of why they are as they are. This latter task is performed by theories, which thereby provide an understanding of the facts, while the former is accomplished through the establishment of empirical laws, the expressions of which are seen to provide scientific knowledge.

The way in which scientific knowledge is provided by laws will be examined in the sections directly following; but first something of an overview may be in order. Here, following Campbell, the set of empirical laws is seen to constitute a highly interconnected network. One law cannot be altered without repercussions being felt throughout the system. Though this system itself is constantly being modified and supplemented, it undergoes no revolutionary changes. Laws considered valid three hundred years ago are still considered valid today, within certain limits – limits which might well have been recognised at the time of the original formulation of the laws. New laws added to the system are generally in keeping with those already present, and are intimately related to them. In this way there is a continuity in the development of the network as a whole; and it is this network that constitutes the empirical aspect of science.

6. EMPIRICAL LAWS AND MEASUREMENT

Apart from principles going under the name of laws (e.g. those of thermodynamics, or those of Newtonian mechanics), virtually every law in physics is dependent for its existence on the results of measurement. Though one perhaps sees a waning in the importance of measurement as one moves away from physics, through chemistry, and on to biology, the central role measurement plays in the paradigm of the sciences makes its consideration a prerequisite to our understanding of science as a whole.

Some caution may be required, however, in speaking of *empirical* laws as dependent on measurement. The term "empirical" has two distinct senses, which ought not be conflated. The one sense may be called the scientist's, and the other, the philosopher's; both are included in the *Concise Oxford Dictionary*. There, the former sense of "empirical" is: 'Based or acting on observation or experiment, not on theory,' while the latter is: 'regarding sense-data as valid information; deriving knowledge from experience alone.' It is in the former and not the latter sense that the laws of the exact sciences are here being termed "empirical," for, as we shall see, one of the primary purposes of measurement in science is to circumvent the errors to which unaided sensory perception may lead.

Empirical laws in the exact sciences are most often expressed as *equations* representing relations between *measurable parameters*. These equations indicate mathematical functions relating quantities of certain properties, quantities which are themselves determined via the performance of specific mensural operations involving the use of particular instruments. The quantifiable properties or parameters appearing in (the expressions of) empirical laws can, following Campbell, be divided into two groups: those which can be measured fundamentally, and those which cannot, the former being parameters which can be measured without measuring something else.[13]

Consideration of what is involved in the performance of fundamental measurement provides a key to the understanding of the nature of objectivity in science. Taking the measurement of length as an example, we see the necessity of there existing a *physical standard* of one unit length, in terms of which all measurements of length must be made. Precisely what entity functions as the standard is not of great moment. But, whether it consist of a particular metal bar or a certain number of wavelengths of a particular colour of light, it ought to be something which is essentially unique and which, as best as we can ascertain, does not change through time. Thus, for example, the scientist does not measure length in 'feet,' where the feet in question are his own.

What has been said about length applies equally well to the other parameters susceptible of fundamental measurement. In each case there must exist an unchanging physical standard with respect to which all measurements of the parameter in question are made. This is the way in which the objectivity of fundamental measurement is ensured. And a moment's reflection reveals this to be the case not only with fundamental measurement, but with *all* measurement.

---

[13] Campbell (1938), p. 127.

Derived measurement also rests on the standards for fundamental measurement, as well as having those of its own; in the simplest case the standard unique to a particular derived measurement may consist solely of a combination of standards for fundamental measurement. But, in any case, we see that any form of measurement must be related to non-changing physical standards, and that it is ultimately by virtue of these that the errors and inaccuracies to which unaided sensory perception is prone are avoided, and mensural objectivity attained.

7. LAWS ARE DISCOVERED; THEORIES ARE CREATED

Since (the expressions of) empirical laws, as conceived here, indicate relations that invariably hold between particular measurable parameters, it is clear that the establishment of a law will itself depend on measurement. The application of laws is thus seen to involve the active performance of specific operations with particular instruments. In order to determine whether a certain relation between a limited number of parameters holds, it is furthermore necessary to exclude the influence of parameters other than those being investigated. This in turn means that the establishment of a law almost invariably requires the creation of a highly artificial, idealised situation, or, in other words, the active controlling of the experimental conditions relevant to the application of the law.

When certain conditions have come to be controlled in a way which, say, had earlier been technologically unfeasible, the scientist is in the position of possibly *discovering* a new law. However, such a discovery does not suffice to answer the question of why the lawfully related parameters are related in the particular way that they are. To answer this question one must turn to the theoretical aspect of science.

Where empirical laws or their expressions are discovered from amongst previously amassed data or as the result of the performance of new experiments, theories are not discovered but are rather constructed or *created* in order to explain why the laws are as they are. By their very nature they are speculative and hypothetical, and are not directly suggested by empirical states of affairs. Technological advance may of course one day lead to the discovery that a particular theory is essentially correct (or mistaken), but at the time when it is first proposed it is nothing more than an intellectual creation of the theoretician.

The kinetic theory of gases, for example, though now known to be correct, was not arrived at as a direct result of the performance of operations in a laboratory situation, but was created by Bernoulli on the basis of reflections upon the then known gas laws. In reaching out into the unknown in order to explain the known, theories demand a transcendence of the facts that is not to be found in the case of empirical laws.

8. CAMPBELL'S CONCEPT OF ANALOGY

In *Physics: The Elements*, Campbell takes up two features of theories that are to distinguish them from experimental laws. One is that a theory should contain propositions referring to hypothetical ideas, such ideas differing from measurable concepts in that they may be attributed infinitesimal values, or have referents which are extremely small or are located in the distant past; and the other is that the propositions of a theory should display an analogy to known laws – either to the laws it explains, in which case the theory is 'mathematical,' or to other laws, whereby it is 'mechanical.'

Campbell lays perhaps more weight on the analogical aspect of theories than on their containing hypothetical ideas, saying that on his view, "analogies are not 'aids' to the establishment of theories; they are an utterly essential part of theories, without which theories would be completely valueless and unworthy of the name."[14] For Campbell, it is in the ideas evoked by analogies that the value of theories lies; and it is when the hypothetical ideas are evoked by the analogy that a theory is truly explanatory.

In considering Campbell's view, one might note the particular way in which he employs the notion of analogy, and the fact that he does not at all use the term "model." Campbell's 'analogy' is one between the 'propositions' in terms of which the theory is expressed and (the expressions of) known laws; i.e. he seems to be concerned either with a sort of isomorphism between expressions, or with an identity

---

[14] Campbell (1920), p. 129.

of form of different states of affairs. Further, he apparently demands only that the analogy be to laws that are known, without requiring that they be laws of any particular kind – though he does go on to suggest that the laws of mechanics, being concerned with matter and force, function best as the known laws when it comes to providing satisfying explanations.

A basic weakness in Campbell's position, however, as seen from the present point of view, is that it does not capture the causal aspect of scientific explanation.[15] Though in his subsequent exposition it becomes clear that Campbell wants his notion of analogy to do more than merely indicate a formal isomorphism, how it is to do this is not clear from what he says in the laying out of his view. To say that the propositions expressing a theory have the same form as known laws may prove an aid to understanding the theory,[16] but it is not necessarily an aid to understanding why the laws the theory is intended to explain take the form that they do. One reason for this is that it does not, in itself, indicate how the laws come to be as they are as a result of the operations of physical entities.

9. MODELS VS. ANALOGUES

In discussing the nature of scientific theory it is important to distinguish analogues relevant to theorising from theoretical models, as these are normally understood in science.[17] The tendency to conflate these two notions has been largely responsible for the widely held view that models play only a heuristic role in science.

Analogues – i.e. that to which analogies are drawn – may appear in various scientific contexts. In Campbell's case, the analogues themselves are to be known laws, and the analogy is to hold between them and the propositions of the theory. They may be called 'formal analogues' (though we note that what we actually have here is not an

---

[15] Cf. *this volume*, pp. 106, n. 25, and 179&n.

[16] As suggested by Hempel: (1965), p. 441.

[17] This distinction may be found in M. Hesse (1963), Ch. 1, and is emphasised in Graves (1971), pp. 48–50. Hempel also makes it in his (1970), p. 157, though he misses it in his (1965); Nagel too misses it in his (1961); cf. pp. 110–114. The distinction is manifest in Harré's work as that between a model and its source, as will be further developed below.

analogy but an identity of form) and may be contrasted with substantial analogues, which may take the form e.g. of gears and pulleys intended to provide a comprehensible, albeit unrealistic, conception of some natural process. In this latter case we should say that the gears-and-pulleys arrangement is analogous to the real process.

Substantial analogues are similar to theoretical models, but an essential difference between the two is that the former are not advanced as serious candidates for the depiction of real states of affairs. In this way, theoretical models are not really analogues at all (though they may function as such subsequent to their original employment), for they are not put forward with the suggestion that they are analogous to certain real states of affairs, but rather with the suggestion that, in causally relevant respects, they might actually depict such states of affairs.

One important function that formal and substantial analogues may have is to suggest the nature of a theoretical model. In such a case the analogy holds between the analogue and the model, the analogue functioning as the intellectual *source* of the model.[18] In this way, for example, the notion of billiard balls colliding with one another and with the cushions of a billiard table may be said to be the substantial source-analogue of the ideal gas model, and Newton's laws of motion may be seen as its formal source-analogue.

It is thus the source-analogue which provides the 'that which is familiar' which suggests the basis of a scientific explanation, the substantial source-analogue suggesting the nature of the causal relation depicted in the model, while the formal source-analogue provides the relation's specific form. The substantial analogue, in order to perform its task, must itself consist of conceivable states of affairs involving physical objects; and it is perhaps the more fundamental analogue of the two.

The above example of the ideal gas model is a particularly straightforward one, and it should be noted that a variety of relations between models and their analogues are possible. For example, some quite acceptable models, pace Campbell, may have no formal analogues, but may function according to laws of their own; or a model may have a number of source-analogues, taken from different fields.[19]

---

[18] This terminology originates with Harré – cf. his (1970a), pp. 38ff, and (1970b), pp. 288ff. – and has been employed earlier in the present volume.
[19] On this latter point, see Harré (1970a), pp. 45–46.

It may be argued that, once a theoretical model has been constructed in analogy to its (substantial) source-analogue, the source-analogue, having played its role in suggesting the model, may be dropped. This would of course correspond to the common view that models or analogues are dispensable. That the substantial source-analogue no longer be required after the model is constructed seems warranted by considerations of simplicity and by the fact that one of the purposes of a theoretical model is precisely to represent those aspects of its analogue which are to be relevant to the eventual explanatory tasks of the theory. Against this, however, it may be argued that the source is nevertheless essential, the reason being that the intuitive content of the model cannot be grasped without it.

10 THEORETICAL MODELS ARE IDEALISED ABSTRACTIONS FROM THEIR SOURCES

On the view taken here, theoretical models constitute the cores of scientific theories. It is models that provide the causal aspect of theories, in virtue of which they have the explanatory power they do. The present view is thus very similar to that of E. H. Hutten, where he says:

Though physicists do not always speak of a model when considering a whole theory, it is obviously not a serious extension of this term to describe the general frame of concepts underlying a theory in this manner. The advantage lies in that the model permits us to bring out the assumptions tacitly presupposed in a scientific theory; for these assumptions are, at best, only vaguely mentioned in the context or perhaps indicated by a diagram, but often they are not even verbally formulated. Without being clear about these assumptions it is impossible, however, to understand the theory, that is, to know how to interpret the formulae.[20]

Where the substantial source-analogue of a model is a relatively complex state of affairs, the model at the heart of a theory is invariably much simpler. It is simpler partly in order to be conceptually (often mathematically) manageable, and partly because not all of the parameters present in the source are of relevance in explaining that which the theory is intended to explain.

---

[20] Hutten (1954), pp. 292–293.

The first aspect of the simplicity of a theoretical model may be expressed by saying that the model is an *idealisation* with respect to its substantial source. It depicts entities which are to behave in strictly lawful ways (perhaps as suggested by the model's formal source), thus greatly assisting in the application of the model to empirical states of affairs. For example, by allowing that the collisions of the molecules represented by the ideal gas model are perfectly elastic, the mathematical description of the model is made much easier, as is the derivation of the empirical laws which are to be explained.

A problem, however, is that while idealisation is necessary in order that the model be understood and its implications determinable, an excess of idealisation can produce a model in which too many causally relevant factors are ignored or over-simplified, resulting in its manifesting itself in empirically unrealistic ways. Thus a balance must be maintained between idealisation and completeness, so that the model be understandable and at the same time sufficiently realistic.

The second aspect mentioned above, that not all of the parameters manifest in the substantial source are represented in the model, is also a form of idealisation; but we may say more particularly that it is in this regard that a theoretical model is an *abstraction* with respect to its source. Thus, while the substantial source of the ideal gas model consists of physical balls knocking against each other, each ball having a particular colour, in the model the notion of colour is completely absent, as it is of no relevance in explaining the gas laws.

A more general use of the term "abstraction" however would suggest that a theoretical model as a whole is obtained by abstraction from its substantial source-analogue. Furthermore, such a model is an idealisation in that it is simpler than its source, and more regular in its behaviour. Thus, at this rather general level, it would not be amiss to characterise theoretical models as idealised abstractions.

11. THEORETICAL ONTOLOGIES AND CAUSAL MECHANISMS

Where empirical laws indicate the specific form taken by the relation amongst various measurable parameters, theories explain that form by indicating its causal basis. This explanation involves the employment of the notion of some kind of substance or matter, between parts of which the causal relation, as conceived in science, can manifest itself. The *general* nature of this matter is determined by the *principles* of the science, while its *particular* nature is determined by *theory*. Thus various theories depict it differently, though each depiction lies within the limits dictated by the constraining principles.[21] In the paradigmatic case, one theory may suggest that the substance is spatially discrete, while another suggests it to be continuous (cf. e.g. the particle-wave dichotomy). Or two theories may agree in depicting the substance as discrete, but differ on how many sorts of it there are, or on what the natures of the sorts are. These different conceptions, as suggested by the preceding section, are obtained by abstraction from their various source-analogues, which may themselves consist of everyday states of affairs involving the interaction of physical objects.

These conceptions of the particular nature of the underlying substance investigated by the science are manifest in the theoretical models at the cores of scientific theories. In this way the theories, or their models, provide *ontologies* which vie for the status of best reflecting the particular nature of that aspect of reality investigated by the science.[22]

Such ontologies have two aspects, one formal and one substantial, corresponding to the theoretical model's formal and substantial source-analogues. Emphasis in contemporary thought has been on the formal aspect of theories, as though a theory could be scientifically adequate so long as it fulfilled certain formal requirements. But this is not so. While the formal aspect of a theory may be necessary for prediction and for the use of the theory in technological contexts, even this cannot be accomplished without the theory's first being linked to empirical laws. And this in turn must involve the theory's substantial aspect, for it is the substantial aspect that indicates how the formalism is to be connected to reality. Thus the substantial aspect is not only necessary for the provision of the notion of cause,

---

[21] On the function of principles as constraints on theorising, see Harré (1970b) and Dilworth (2007).

[22] For a discussion of the role of models in forming ontological hypotheses, see Graves (1971), pp. 53ff.

without which the theory cannot function in an explanatory capacity, but it is necessary for its very application to empirical reality.

In order for the notion of cause to emanate from a theory, the model underlying the theory cannot be static but must depict moving entities which causally interact. In other words, in order for the theory to be explanatory, its ontology must include *causal mechanisms*. Now some mechanisms depicted by the theories of current science have rather peculiar properties, distinguishing them from the ideal sketched earlier in more ways than by their simply including action at a distance. The history of science, or more particularly of physics, has been one of an increasing liberality as regards the allowable forms of theoretical ontologies. This tolerance has occurred in response to the unwillingness of nature to adapt to the original ideal, and might well make possible the ultimate erection of a new ideal. But at which point the methodology of theoretical science will have diverged too far from the ideal sketched above in order still to warrant the appellation "scientific," or at which point we might want to say that the term "science" has a basically different meaning from what it has, say, in reference to the Scientific Revolution, are not issues which will be pursued here. What is being asserted here, on the other hand, is that the essence of explanation as provided by scientific theory, through all its changes to date, still consists in the depiction of some sort of causal mechanism conceived as responsible for the manifestation of the facts or laws the theory is intended to explain.

Now it is most frequently the case that the mechanisms deemed responsible for known empirical regularities are not susceptible of direct observation. As intimated earlier, if they were, the cause of the regularities would be immediately evident, and no theorising would be required. This is what gives scientific theory its theoretical aspect – models depicting causal mechanisms are arrived at on the basis of informed speculation, and are originally put forward only hypothetically.

The general approach taken here with respect to scientific models, explanations and causal mechanisms has also been taken by others, among them Alexander, where he says:

However more complex and mathematical the sciences have become, they rely constantly upon hidden mechanisms. One of the things which makes for greater complexity is that the mechanisms are more securely hidden; another is that the mechanisms may not always be 'mechanistic' in the sense of classical mechanics. But the principle is the same. Scientific explanations mention something going on in terms of which observable events can be understood. Models ... can ... be regarded as attempted descriptions of these hidden mechanisms.[23]

12. THE NOMINAL VS. THE REAL ASPECT OF THE DOMAIN OF A THEORY

Just as a theoretical model or theory ought to be distinguished on the one side from its origin – its source-analogue(s) – it ought also be distinguished on the other side from that to which it is applied. The latter will here be termed the theory's *domain*.

This notion allows of a further distinction between what may be called the real and the nominal aspects of the domain. The domain's *real* aspect may be characterised as that which the theory represents, while its *nominal* aspect may be depicted as that which the theory is intended to explain. The real aspect of the domain of a theory should thus be responsible for the manifestation of the particular known facts or laws to be explained. It is thus represented in the theory as consisting of certain entities which constitute causal mechanisms whose functioning is intended to manifest itself in the empirical laws in question. Depending on how successful it is in achieving this end, the theory may be said to explain these laws.

The nominal aspect of the domain, on the other hand, consists of these empirical facts or laws themselves, or of the data accumulated in their registration; i.e. it consists of the overt behaviour of that which is being studied. Thus the nominal aspect of the domain of the kinetic theory of gases, for example, consists of the empirical laws involving pressure and temperature (the known gas laws) which the theory is intended to explain, while the real aspect is the gas itself, conceived as consisting of multitudes of tiny particles in rapid motion.

On deductivist approaches a fundamental problem arises regarding how theoretical propositions, which refer to unobservables (the domain's real aspect), are deductively to subsume propositions

---

[23] Alexander (1963), p. 137.

expressing empirical laws (the domain's nominal aspect), which refer only to observables. Here this problem is avoided, for the formal relation of interest is not conceived as involving the deduction of one set of *propositions* from another, but the mathematical derivation of one set of *equations* from another. The two sets of equations are, respectively, the one expressing the formalism of the theory – i.e. the theory's formal aspect – and the one expressing the empirical laws constituting the nominal aspect of the theory's domain. The former are laws intended to apply to the causal mechanisms which are the theoretical entities the theory depicts: they might consequently be called theoretical laws.[24] They differ however from what Nagel calls theoretical laws in that they do not alone constitute the theory, nor even its most important aspect, and in that they are not propositions about unobservables, but equations whose variables represent properties of unobservables.

Whether theoretical laws apply to reality is an open question so long as the entities depicted in the theory are merely hypothetical; but they may be said to apply to the constituents of the theoretical model, i.e. to the representation of reality, and to that exactly. Theoretical laws thus express the form or structure of the causal relations depicted in the theory, but not their actual causal or substantial aspects. In so doing they contain variables standing for properties and relations of the theory's theoretical entities.

The *application* of the theory consists in the construction of a particular mapping operation from the values of variables in the theoretical laws to values of the parameters in the empirical law(s) to be explained. This mapping is performed in accordance with physical principles, such as that of the conservation of energy, and is done in such a way that all empirical parameters find expression in substantial and/or causal notions. In the theory of gases, for example, the values of variables standing for the *velocities of the molecules* of a gas are mapped onto the value of the parameter *temperature* in the equation expressing the particular empirical law the theory is being used to explain. Here the mapping itself consists of the function expressing the mean kinetic energy of the molecules of the gas.

---

[24] This usage of the term is narrower than that in the main text of the present volume, where theoretical laws are taken to include other laws derivable from the theory.

The mathematical 'derivation' of empirical laws from theoretical laws does not alone constitute the bridge between theories and laws however, for the derivation would not even make sense were it not intended that each sort of law concern (a different aspect of) the *same* reality, and that the causal basis of that reality rest in the aspect depicted by the *theory* and not by the empirical laws. Theoretical laws, in that they express the form of causal processes manifest in the substantial aspect of the theory, thus serve as *part* of a chain the whole of which *explains* the empirical laws by showing them to have a causal basis. And it is in this way that the nature of the *nominal* aspect of the domain of a theory is shown to result from the nature of the domain's *real* aspect.

The conception provided above as to how empirical laws are derived from theories is but one instance of how we are here capable of solving problems of current philosophy of science, the problem in this case being that of correspondence rules. The problem of the purpose of theoretical terms is also solved here, for they may be seen as terms referring to the substantial entities depicted in theories – entities necessary to the provision of a causal explanation of empirical laws. Further, the nature of theory change is also explained: one ontology representing physical reality is replaced by another, thus giving theory change a revolutionary aspect, while at the same time the empirical laws to be explained by the respective theories remain relatively unaltered, providing continuity in the development of science.

More generally, the advantages of the present approach over formalistic approaches in the philosophy of science lie in its comprehensiveness and, more particularly, in its explicit recognition of the non-formal aspects of science. Natural science is not logic, nor is it mathematics; and though the latter plays a vital role both in the expression of empirical laws and in their derivation from theories, it does not suffice to capture the causal aspect of theories nor the mensural nature of laws. Physical theories, in order to fulfil their explanatory function, must involve the conception of causal relations between material entities; and empirical laws, in order to provide physical knowledge, must relate parameters the values of which are determinable through the appropriate use of measuring instruments. It has been the task of this appendix to show how and why this is so.

IS THE TRANSITION FROM ABSOLUTE TO RELATIVE SPACE A SHIFT OF CONCEPTUAL PERSPECTIVE?

In the present volume the attempt is made to provide a conception of science capable of accounting for both its revolutionary aspect and its apparent progress. On this view – the Perspectivist conception – a distinction is drawn between *laws* and *theories*, laws being seen as providing the empirical aspect of science, which theories in turn are constructed to explain. Since the writing of the book, however, a third fundamental category of scientific thought has presented itself, namely that of *principles*.

One way of conceiving of principles is as setting constraints on the form that laws and theories can take. In this way they are more fundamental to a scientific discipline than are laws and theories, and, in effect, they actually serve to define it. A change of principles is not impossible however, but such a change is more drastic than is a change of theory, and brings with it a change in the nature of the discipline itself.[1]

In this volume the Perspectivist conception has been developed in application to theories, each theory being seen to constitute a *conceptual perspective*. Now, with the subsequent entrance of principles into the discussion, the question arises as to the extent to which principles, or groups of them, can also be seen to have the characteristics requisite for constituting conceptual perspectives. This question is of direct relevance to that posed in the title of this appendix, for it would appear that questions concerning the nature of physical space are more matters of principle than of theory; and so the present topic may provide a context in which the more general question can be broached.

---

[1] These matters are treated in much greater detail in Dilworth (2007).

The issue regarding absolute and relative space is taken up by Newton in his *Principia*, where he argues for the existence of absolute space, saying that "The effects which distinguish absolute from relative motion are the forces of receding from the axis of circular motion."[2] He goes on to describe the famous bucket experiment, as well as that of the two globes: "For instance, if two globes, kept at a given distance one from the other by means of a cord that connects them, were revolved about their common centre of gravity, we might, from the tension of the cord, discover the endeavor of the globes to recede from the axis of their motion, and from thence we might compute the quantity of their circular motions."

The classic rebuttal of Newton's argument is provided by Mach, and is concisely expressed where he says: "Try to fix Newton's bucket [or his globes] and rotate the heaven of fixed stars and then prove the absence of centrifugal forces."[3] In other words, the centrifugal forces which Newton takes to result from motion relative to absolute space might just as well result from motion relative to other masses.

Does the move from one to the other of these two points of view constitute a shift of conceptual perspective? To determine whether it does, we should consider it in terms of the five notions central to the Perspectivist conception.

1. INTENDED DOMAIN

The first question to be asked with regard to this notion in the case of perspectival shifts is whether the intended domains of the relevant perspectives are or are not the same. In the paradigmatic case, and in order to be susceptible of evaluation in terms of the criteria to be given below, they should be the same. As regards the move from the Newtonian to the Machian conception, this appears to cause no difficulty, both views quite clearly being applied to the same realm.

Just what this realm is, however, is more difficult to say. It would seem at first that it ought to be physical space, but further consideration suggests that it is rather the behaviour of physical objects in

---

[2] Newton (1687), p. 10; next quote, p. 12.
[3] Mach (1883), p. 279.

space. This behaviour is of a particular sort, i.e. that dependent upon the effects of inertia, Newton characterising these effects as stemming from the nature of (absolute) space, while Mach suggests them to arise as a consequence of the presence of other masses.

It is of some interest to note that if we were to take the common intended domain here to be physical space – i.e. if we were to consider only what Newton and Mach have to say about space – then we would not have a shift of conceptual perspective. This is so because each of two competing perspectives must offer a *positive* characterisation of their common domain,[4] while here we would have Newton characterising (absolute) space as providing the causal basis of inertial effects and Mach simply denying that space has this capacity. To have a perspectival shift we have to include Mach's positive thesis that inertial effects are dependent on the presence of other masses.

2. SAMENESS OF CATEGORY

Another requirement of perspectival shifts, in order that they be of the sort where the question of competition arises, is that the perspectives concerned involve the employment of predicates of one and the same category. When the perspectives are applied *theories*, the categories which are of relevance are measurable parameters, and evidence suggesting the particular parameters with which a theory is involved may be obtained through considering the sorts of operations or measurements used in applying or testing the theory. Thus two distinct theories' pertaining to the same measurements suggests that they involve the same categories or parameters.

In the present case too then we might look at the measurements or operations involved in the application of the Newtonian and Machian conceptions, to see if they suggest certain categories or parameters as being relevant to both. Here the paradigm operation is that of rotating a bucket or two globes – and with this we should include Mach's thought experiment of rotating the heavens themselves – and subsequently determining what results from the rotation. In each case we are considering the application of a force (to set the relevant body or bodies in motion), and, more importantly, the measurement of a

---

[4] Cf. *this volume*, pp. 71–74.

force (that resulting from the motion). So it would appear that a category fundamental to the application of both conceptions is that of force, i.e. that the two views have this parameter in common. And the fact that the same operations are relevant to the application of both of them reinforces the view that, considered as conceptual perspectives, they share a particular intended domain.

3. PERSPECTIVAL INCOMPATIBILITY

Perspectival incompatibility, which arises when different perspectives characterise a common domain in mutually exclusive ways, is to be distinguished from an 'incompatibility of results,' which concerns the ways in which the perspectives are manifest e.g. in instrument readings. While an incompatibility of results may imply an underlying perspectival incompatibility, the reverse need not be the case.

Viewed as conceptual perspectives, are Newton's and Mach's respective conceptions of absolute and relative space perspectivally incompatible? Mach himself argues that Newton did not need the notion of absolute space for his mechanics,[5] since, in order to have a generally valid system of reference, he in effect chose the one determined by the fixed stars.[6]

This is not an argument, however, for the compatibility of the two views – nor does Mach intend it as such. What Mach is saying is that Newton's actual choice of an 'absolute' system of reference (inertial system) was not determined with respect to space itself but with respect to bodies in space, i.e. it was, in Mach's terms, relatively determined. As it turns out, it was also determined in such a way as to be coincident with Mach's subsequent 'relative' system of reference, both systems being seen as stationary in relation to the fixed stars.

Did these two systems – these two spaces – have to be coincident? Mach sees relative space as determined by the distribution of matter in the universe, and Newton sees absolute space as determining inertial systems. Taking the matter of the universe as mainly residing in

---

[5] For the contrary view, see e.g. Einstein (1952), p. 135.
[6] Mach (1883), p. 285.

its stars, and the stars as inertial systems, then the two spaces at least ought not be accelerating with respect to one another. But even admitting that they are relatively stationary, should this mean that the perspectives in which they are respectively framed are compatible?

Here we have an instance of perspectival incompatibility in spite of what, at least on first consideration, may be considered a compatibility of results. On Mach's view, inertial systems are determined by matter; on Newton's view, by (absolute) space. This alone is sufficient for their being perspectivally incompatible. In a similar vein we could consider a view in which the nature of absolute space determines not only inertial systems but also the distribution of matter in the universe. Given such a distribution a spatial relativist might conclude, to the contrary, that it is the distribution of matter which determines the inertial systems. Again we would have an instance of perspectival incompatibility between the views in question. In both of the above cases states of affairs are depicted which cannot coexist, even if each should manifest itself in the same way as the other.

If matter is not accelerating with respect to Newton's absolute space, then the inertial systems on Mach's and Newton's conceptions are the same. But there is nothing in principle to preclude the vast majority of the matter in the universe from undergoing a uniform acceleration with respect to absolute space, while this is impossible with respect to Mach's relative space. This too shows the two views to be perspectivally incompatible.

4. LOGICAL SIMULTANEITY

That two conceptions are perspectivally incompatible implies that they cannot be applied *simultaneously* to the same intended domain, independently of whether their subsequent characterisation of that domain is the same. In the present case this should mean that while bodies inertially affected by their state of motion may be conceived in either a Newtonian or a Machian way, they cannot be conceived in both ways at the same time, even if the fixed stars are taken to be at rest relative to absolute space.

This notion of simultaneity – logical simultaneity – is intimately related to the notion of perspectival incompatibility, the reason that the Machian and Newtonian conceptions cannot be applied logically simultaneously also being the reason that they are perspectivally incompatible: they characterise one and the same state of affairs in mutually exclusive ways, even if those ways are such that the difference between them should never be empirically manifest. Those systems of coordinates which constitute inertial frames cannot be conceived simultaneously as being determined both by space itself and by the matter distribution of the universe, independently of whether the same systems of coordinates are determined in both cases.

5. RELATIVE ACCEPTABILITY

It need not always be the case that incompatible perspectives can be judged as to their relative acceptability, but when such a judgement is possible it may be seen as being made in terms of some combination of the relative accuracy, scope and simplicity of the perspectives.

Can either the Newtonian or the Machian conception be said to be more *accurate* than the other? If we assume the fixed stars to be motionless in absolute space, it would seem that such a question ought not arise, since the inertial effects of motion on either conception should manifest themselves in the same way as on the other. However, taking the Newtonian and Machian views to concern the cause of inertial phenomena, we may say that Newton attributes that cause to (absolute) space, while Mach attributes it to matter. It should thus be in principle possible to conduct an experiment in which a large mass is rotated very fast to see if it produces a centrifugal field (near its axis of rotation) which affects bodies not participating in the rotation. Were it to do so in a detectable way, we should say that Mach's conception is more accurate than Newton's – this small difference in accuracy reflecting the fundamental untenability of the Newtonian view and so making the experiment a crucial one.[7]

Had the effect been clearly shown to result from the relative rotation of the mass in question, we should say that the *scope* of Mach's

---

[7] Such an experiment has in fact been performed, producing the effect expected on the Machian conception, though it may have been due to other causes: cf. Reichenbach (1928), pp. 214–215&n.

view is greater than that of Newton's. While both views apply equally well to normally produced centrifugal forces, Mach's view would apply to those produced in the above manner as well.

The *simpler* of two perspectives is the one requiring the fewer ad hoc modifications in order to apply to its intended domain. Thus, for the sake of illustration, we might imagine a Newtonian who, faced with results favouring Mach in the above experiment, claimed that inertial frames were normally determined by (absolute) space, but that they could also be determined by masses in motion with respect to space. Were the Newtonian unable to show why this should be so (e.g. as the result of some general principle encompassing both sorts of case) we should say that his emendation was merely ad hoc, and that his subsequent (revised) view was accordingly less simple than the more coherent view of Mach.

The above considerations suggest that the Perspectivist conception might be applied with some profit to the respective views of Newton and Mach. Each view constitutes a conceptual system providing a structure in which one's thoughts about some aspect of the world can be organised, as may be expected of a conceptual perspective; and the move from the one to the other involves a fundamental change in one's way of conceiving of the basic principles of mechanics, a change which can be well characterised in terms of the basic notions of the Perspectivist view. And this in turn suggests that scientific principles more generally might also be susceptible of a Perspectivist interpretation.

TWO PERSPECTIVES ON SUSTAINABLE DEVELOPMENT

The primary goal of the leaders of virtually all countries in the world today is to increase the rate of economic growth in their respective nations. And though it has come to be recognised by virtually everyone that a healthy economy ultimately depends on a healthy environment, and that the state of the environment throughout the world is rapidly deteriorating, a glance at the issues emphasised in the media indicates that most people share the priorities of their leaders. While it is admitted that steps must be taken to improve the state of the environment, such problems as increasing unemployment and falling real wages are felt to be much more pressing, and their solution is sought in the new jobs and greater income expected to accompany increased production.

That economic growth should be highly valued is of course not a new phenomenon, being part of the Western ethos at least since the time of the industrial revolution. But its continued pursuit without regard for the environment is being viewed with increasing consternation by those concerned about the future well-being of humanity. The document expressing this consternation that has received the most attention in recent years is the report prepared by the United Nations' World Commission on Environment and Development, chaired by Gro Harlem Brundtland. The key concept to emerge from the report, entitled *Our Common Future*, is that of *sustainable development*; and in the report the plea is made to the world at large, and to politicians, businessmen and bureaucrats in particular, to ensure that future development does not jeopardise the environment on which we all depend, i.e. that it be 'sustainable.'

As presented by the Brundtland team, development is sustainable if it satisfies present needs without compromising the ability of future generations to meet their needs.[1] While this definition could undoubtedly stand some clarification with regard to the notion of sustainability, our attention is drawn rather to the fact that it gives no idea of what *development* is to consist in. The content of *Our Common Future* on the other hand makes it clear that by development is intended nothing other than *economic growth*, and moreover that the Brundtland team encourages such growth as an aid both in overcoming the world's environmental problems as well as in improving the lot of the world's poor:

Far from requiring the cessation of economic growth, [sustainable development] recognises that the problems of poverty and underdevelopment cannot be solved unless we have a new era of growth in which developing countries play a large role and reap large benefits. ... If large parts of the developing world are to avert economic, social, and environmental catastrophes, it is essential that global economic growth be revitalised. In practical terms this means more rapid economic growth in both industrial and developing countries.

The secretary general of the Commission, Jim MacNeill, is even more specific regarding:

what we called the *growth imperative*. The world's economy must grow and grow fast if it is to meet the needs and aspirations of present and future generations. The Commission estimated that a five- to ten-fold increase in economic activity would be needed over the next half century just to raise consumption in developing countries to more equitable levels. Energy use alone would have to increase by a factor of eight, just to bring developing countries, with their present populations, up to the level now prevailing in the industrial world. I could cite similar factors for food, water, shelter, and the other essentials of life.[2]

There has been a growing reaction against this approach however on the part of environmentalists and environmental researchers. For example, AnnMari Jansson of the Department of Systems Ecology at the University of Stockholm has said:

Facing the enormous task of stopping further environmental degradation, it is hard to understand the assessment of the Brundtland Commission that the

---

[1] World Commission on Environment and Development (1987), p. 8. Quote following, ibid., pp. 40, 89; cf. also pp. 50–51, 173, 335.
[2] MacNeill (1989), p. 18.

international economy must speed up world growth. Rather, we need to adopt a strategy to reach sustainability through a transition from economic growth based on the depletion of resources towards a new progress based on environmental management. The rationale of market economies enforces competition and profit maximisation, putting a premium on short-run productivity to such an extent that the long-run carrying capacity of, e.g., agricultural ecosystems is destroyed. We need to replace this short-run rationale with environmental management that fully considers the economic importance of nature's capacity to serve human society.[3]

And to this may be added Herman Daly's comment, that "The growth ideology is extremely attractive politically because it offers a solution to poverty without requiring the moral disciplines of sharing and population control."

One way of characterising these two views rather generally is to say that the approach advocated by the Brundtland team (which for ease of reference we shall term the *economic* approach) sees economic growth – promoting the development of environmentally friendly technology – as a prerequisite for the attaining of ecological sustainability, while the other approach (the *ecological*) sees sustainability as the immediate goal, the attaining of which is only hindered by the striving for economic growth.

A distinction along these lines has been recognised by a number of authors, and in some cases it has been suggested that the alternative views are in fact widely held, and that at the same time the difference between them cuts deep. In this regard it may be asked, for example, whether the 'economic' approach of the Brundtland team is not just an expression of the widespread Western view of economics and the environment cited at the beginning of this appendix, albeit one in which greater emphasis is placed on the environment, while the 'ecological' approach suggests a radically new way of viewing the situation. It may also be wondered whether the respective views are limited to just economics and the environment, or whether they are expressions of alternative ways of viewing reality as a whole, that is, whether they constitute alternative *worldviews*.

In the present appendix an analysis will be made of the 'economic' and 'ecological' approaches to sustainable development in the context of the structure provided by the Perspectivist conception.

---

[3] Jansson (1988), p. 35; next quote, Daly (1990), p. 242.

Thus each of these views will be considered to constitute a 'perspective' in the technical sense, and it will be seen what light can be shed on the issue by so viewing them. That some success might be had is suggested e.g. by Daly's remark that "To a large degree, the growth debate involves a paradigm shift or a gestalt switch,"[4] as well as by the content of Mary E. Clark's article, 'Rethinking Ecological and Economic Education: A Gestalt Shift.'[5]

The potential success of this approach is strengthened by the fact that perspectives constitute conceptual systems or frameworks which provide structures in which one's thoughts about some particular aspect of the world can be organised.[6] Thus perspectives are not simply sets of statements which may be true or false, but are ways of conceiving, which may have greater or less applicability. This is clearly similar to the situation involving the economic and ecological approaches to sustainable development, for this case too involves alternative ways of conceiving or making sense of a particular aspect of reality, ways which in themselves need be neither true nor false, while nevertheless being relatively more or less applicable to the phenomena they have in common.

It should be noted however that there are also important differences between the present case and that involving scientific theories, including that the economic and ecological views might well contain *values* as essential elements, that they do not simply concern what is the case in reality but how to act with respect to reality, that neither is so clearly delineated as are scientific theories, and that consequently some of their implications can only be surmised. To a large extent, however, these differences can be accommodated by the Perspectivist conception, as will be seen below.

Here then we shall analyse the relation between the economic and ecological approaches to sustainable development in terms of the five basic factors of the Perspectivist conception.

----

[4] Daly (1992), p. 126; cf. also Daly & Cobb (1989), pp. 5–6. Reference to the present distinction in terms of a paradigm shift was also widespread at the 1992 meeting of the International Society for Ecological Economics.
[5] Clark (1991).
[6] Cf. *this volume*, p. 69.

1. THE INTENDED DOMAIN OF APPLICATION

To what is it intended that the economic and ecological views be applied, and is it the same in both cases? Or have they different 'intended domains of application,' or perhaps not identical but overlapping domains?

It would seem that the economic and ecological perspectives clearly have one point of contact in that both concern sustainable development. But the situation is more complicated than in the case of scientific perspectives. That each perspective involves conceiving of sustainable development differently is not a problem, for scientific perspectives also provide alternative conceptions of their common subject-matter; and such conceptions, though different, may also be expressed using the same or similar terms. But what in the present case is the common reality to which the different conceptions of sustainable development are to be applied; and how can it be characterised in a way that is independent of both of them?

Intuitively we should say that at least part of what both the economic and ecological views are to be applied to is the human economy. But here we see a basic difference between the present case and the scientific one, which lies in its *prescriptive* nature. Scientific perspectives are concerned to provide *understanding* of a particular aspect of reality which leaves that reality *unchanged*, whereas here both perspectives are concerned to suggest *ways of acting* that will lead to a *change* of reality – i.e. to new ways of conducting economic activity such that it becomes 'sustainable.' At the same time, however, it is one and the same economy that both views address, the potential difference resting in the form that that economy is to take given the application of each of them. Thus we can characterise at least part of the intended domain of the two perspectives as consisting in the transfer of material goods and the provision of services – both in exchange for money – such activities constituting a 'market.'

But then the question arises as to whether the two views have more in common than this, or whether one of them is intended to apply to more than the other. In this latter regard we note that a number of the advocates of the ecological approach view the economy in such a way as includes what may be termed the 'informal' or 'non-market' economy.[7] A key aspect of the informal economy is that it does not involve money and so does not affect the GNP of a state; thus such activities as simple barter, home gardening and unpaid child-care are considered part of the economy in this broader sense.

The existence of this difference provides a hint that there may be a more fundamental divergence of orientation separating the two views, which at least in the case of the ecological perspective may warrant our thinking in terms of a particular worldview.[8] One way of approaching such a difference may be to suggest that where the economic perspective emphasises the developmental aspect of sustainable development, the ecological perspective emphasises the sustainable aspect. But this is not all, for a perusal of the literature supporting the ecological view indicates that while the attaining of sustainability is accepted as a prerequisite of human activity, many non-economic facets of that activity go into the composition of the complete picture, including e.g. demographic and political decentralisation, regional self-sufficiency and civil as contrasted with military defence, in conjunction with a particular regard for questions of ethics and the idea of community. And it is intended that the changes advocated will have positive effects not only on the environment but on society as well.

Thus it seems that at least in the case of the ecological perspective we may be dealing with a complete worldview, the intended domain of which is human activity generally. Can the same be said of the economic perspective? To what extent is the current worldview in the West shaped by economic considerations? A preliminary answer might be that economics (in the narrower sense) is indeed central to the Western worldview, as is evidenced by the fact that the size of one's income and what one can buy with it play a fundamental role

---

[7] In this regard, cf. Daly & Cobb (1989), Ch. 7. As expressed by Clark (1991, p. 404), with reference to H. Henderson and M. Waring, "the market economy – which includes the so-called 'global economy' – comprises only a small portion of the goods and services upon which we depend."

[8] That this is a correct interpretation is supported e.g. by the content of Daly & Cobb (1989) and Clark (1989), as well as by recent literature produced by the Green movement: cf. e.g. Gahrton (1988) and Moon ed. (1989).

in such activities as voting, choice and pursuit of career, and even everyday conversation.[9]

But is this orientation towards economics only part of a broader perspective, a perspective which might be called materialism, which had its birth with the Presocratic physicalist philosophers and is manifest most clearly in the form of modern science?[10] Thinking along these lines we might distinguish different potential worldviews on the basis of what they consider to be fundamental to reality, whether it be e.g. inert matter, life, or the spiritual realm. These thoughts are rather speculative however, and for the present we may rest content with suggesting that an orientation towards market economics, if not constituting the heart of the Western worldview, is at least an integral part of it, having a strong influence on how people act in their daily lives. As a consequence of this we shall thus say that the economic and ecological perspectives themselves each have as their intended domain of application the world of human activity quite generally.

2. SIMULTANEITY AND SHIFT OF PERSPECTIVE

Though different perspectives can have the same intended domain, there is nevertheless a fundamental sense in which they are independent of one another. Just as the different aspects of the Gestalt Model cannot be seen at one and the same time by the same person, more complex perspectives cannot apply to their common domain simultaneously.[11] They exclude each other in such a way that neither

---

[9] As expressed by Keynes: "[T]he ideas of economists and political philosophers, both when they are right and when they are wrong, are more powerful than is commonly understood. Indeed the world is ruled by little else. Practical men, who believe themselves to be quite exempt from any intellectual influences, are usually the slaves of some defunct economist." (Cited in Clark, 1991, p. 412.)

[10] von Wright speaks of "the worldview modern science has given us" (1986, p. 9). For a presentation of what are suggested to constitute the fundamental principles of modern science, and an explanation of how it has been shaped by them, see Dilworth (2007).

[11] In a context similar to the present one, but where the concern is with alternative worldviews within economics, Benjamin Ward comments: "Several years ago, as I began trying to put together these characterizations of economic world views, an annoying problem began to emerge. I found that it was not possible to work simultaneously on more than one of them." (1979, p. 461).

can logically subsume nor contradict the other; nor can they both be parts of some larger, third perspective. Each constitutes an applied system of thought – and in the present case, of valuation – complete within itself, a system which cannot be intermeshed with the other in any coherent way.

One of the implications of this state of affairs is that it can be difficult for a person to 'see' the world from an alternative perspective for the first time. This cannot be accomplished simply by following a certain number of steps, but requires adopting an uncritical attitude, or having an open mind. The problem here is that there is frequently a strong resistance to adopting such an attitude, one's immediate inclination being rather to 'interpret' all expressions of the alternative perspective in terms of one's own, where one feels secure.[12] In the event that two parties who have different perspectives are both in this situation, a dialogue between them might well be characterised as a *talking at cross purposes*.[13] The same words used by both parties may have essentially different meanings for each of them. For example, a term such as "economic reality," when used by someone speaking from the ecological perspective, could be understood in a way not at all intended, when heard by a person operating from the economic perspective. The remedy for such misunderstandings does not consist in their being removed piecemeal – the attempt at which might only increase the overall misunderstanding – but in helping one's interlocutor make the shift to one's own perspective. Ultimately it is not the meaning of individual words which is to be understood, but a whole way of conceiving of and valuing the world.

It appears that, as a matter of fact, it is more difficult for most people to shift from the economic to the ecological perspective than vice versa. This is understandable since the basic structure of the economic perspective has most often been adopted without reflection in the course of one's being raised in a Western culture, while the

---

[12] In this regard cf. Clark (1992), where she speaks of "the difficulty of getting outside one's own mental framework. [B]ecause we each depend upon and must trust our own worldview to survive, we invest it, and the underlying beliefs on which it is founded, with our emotions. Being truly 'open minded' is thus an uncomfortable – often psychically dangerous – undertaking."

[13] This phenomenon was also frequently referred to at the 1992 International Society of Environmental Economics meeting in Stockholm.

ecological perspective, for those who have it, has usually been arrived at afterwards. Having experienced both perspectives, however, it is not difficult to shift from one to the other, just as one can shift back and forth between the different aspects of the Gestalt Model once one has seen both.[14]

The making of such a shift does not imply that one need accept the newly adopted perspective as superior. Thus, for example, just as where in the scientific case one can shift to an alternative perspective without adopting the beliefs it suggests, here one can 'understand' the values inherent in the other perspective without acting on them.

Virtually all of the leading politicians in the West share a resistance to shifting to the ecological perspective, as is evident from their generally failing to acknowledge its existence, or downplaying it in various ways. The latter consists in their taking it to have only to do with the environment, and in referring to it in pejorative and misleading terms as being, e.g., 'conventional wisdom' and at the same time 'pessimistic.' One striking expression of this attitude outside of the political realm was the claim made recently by a leading philosopher of economics at a conference near Stockholm, that we need not worry about the well-being of future generations because, after all, they will have *more money*!

For those who have made the shift to the ecological perspective and believe it to afford a superior approach to sustainable development, the question of how to get other people to make the shift becomes of paramount importance. Of course this is only a preliminary to getting them to agree that the ecological perspective is in fact superior, but practically speaking it is a necessary preliminary. Direct argumentation from one's own perspective will not work, so one has either to show others that their perspective is incoherent or otherwise unacceptable by standards that it recognises, and so open their minds to the possibility of adopting an alternative, or present the ecological

---

[14] Cf. again Ward: "As each of the world views became internalized and my efforts turned from construction to revision, I found that it became increasingly easy for me to switch from one to the other." Nevertheless, "each switch was total, in that the intuitional elements and the emotional tone of my approach to problems switched at the same time." (1979, p. 461).

perspective in such a manner that it entices of its own accord.[15] In this latter regard there may be much to be said for the use of propaganda, first to create awareness of the existence of the alternative, and second to arouse interest in it.

3. PERSPECTIVAL INCOMPATIBILITY

Perspectives constitute ways of conceiving of – and in the present case valuing – reality, in which the relevant concepts and values are unified into a whole. This synthetic nature of perspectives means that it only makes sense to deal with them as wholes, a point which comes particularly to the fore in the case of comparing competing perspectives.

On the Perspectivist conception, the essential conflict between competing perspectives consists in the fact that the very notions (and/or values) they involve preclude their being applied simultaneously to one and the same realm by the same person, independently of whether their respective application should lead to the same results. An understanding of this perspectival incompatibility in the present case thus requires a comparison of the concepts and values basic to the economic and ecological perspectives respectively.

The *concepts* at the core of the economic perspective are those of individuals (human beings or groups of human beings) exchanging property (goods or services), such exchanges together constituting a market. The primary values are the benefits individuals perceive themselves as capable of deriving through an exchange (depending on their preferences); and each action on the market is conceived as being directed to the goal of obtaining a maximum of such benefits for oneself, such actions being termed rational (the principle of rationality). The market thus constitutes a system of individuals each of whom is acting of their own free will so as to maximise their own benefits, and is furthermore so conceived that in the (ideal) case where all parties succeed in doing so (where supply equals demand), it is in a state of equilibrium (the principle of equilibrium).

---

[15] Cf. Daly, who suggests in a similar vein: "Conversion cannot be logically forced by airtight analytical demonstrations by either side, although dialectical arguments can sharpen the basic issues." (1992, p. 126).

On the economic perspective as applied to the real world, it is precisely through unrestricted rational action that the ideal of equilibrium can be most closely approached; in this way individuals' freedom to act also becomes a value on this perspective, benefiting both the individual actors and the market as a whole. Equilibrium is thus to be approached in practice through real (oligopolistic)[16] competition, in which traders in the same or similar goods or services vie with one another to obtain other goods or services through transactions which result in a maximum of benefit to themselves. To attain this (rational) end they offer their own goods and services at rates of exchange which maximise the benefits to those with whom they trade. In this way then a free market is thought constantly to tend (via 'the invisible hand') to a state of equilibrium. Furthermore, on this perspective it is believed that the more that is exchanged through such transactions, the greater the benefit to the individuals who partake in them. Thus an increase in the production (via technology) and subsequent exchange of goods is considered desirable for all concerned.

Another key notion on the economic perspective is that of money. Money constitutes a quantified socially invested potential or power to acquire property, and the concept of money is employed in the economic conception by defining all market transactions as consisting in the exchange of property for money. The acquisition of money thus constitutes a middle step in the acquisition of property which allows the benefits obtained through exchanges on the market to be accumulated. The quantitative and ubiquitous nature of money also means that it can provide unambiguous information regarding the (marginal) benefits represented by various properties (their prices), thereby facilitating rational action on the market. In these terms, then, the amount of money that changes hands constitutes a measure of economic activity (in the case of a state, its GNP), a greater amount indicating a greater benefit to all.

---

[16] In economic theory, the basic notion of competition is applied to markets in which no individual's actions can affect the bargaining power of other individuals. As is pointed out in elementary textbooks, however, this theoretical notion of 'competition' is virtually the antithesis of the everyday notion. In any case, as regards the real world it is competition in the sense depicted above that is advocated by adherents of the economic perspective.

Where the core of the economic perspective focuses on individuals (human beings) and how they might mutually benefit through
the voluntary exchange of property, that of the ecological perspective
focuses on the whole of the biosphere and how its well-being –
including that of humans – might be maintained. Thus as regards its
conceptual aspect the notion of an (ecological) system is the starting
point of reflections within the ecological perspective, while the
primary *value* is the harmonious working of this system. So where
the conceptual core of the economic perspective is atomistic (the
notion of a system being derived from that of the action of individuals), that of the ecological perspective is holistic (the notion of a
system being primary). And the same may be said as regards their
basic values, the economic perspective seeing an increase in benefits
to individuals as being primary, while the ecological perspective
places value first on the harmony of the whole.

This atomistic/holistic difference also carries over to how *human
action* is perceived on the two perspectives. Where the economic
perspective concentrates on action directed towards specific well-
defined goals, the ecological perspective concentrates on action
directed towards general ill-defined goals.[17] A further difference in
this regard is that the core of the ecological view, rather than dealing
only with interaction between human beings, considers interaction
between humans and their physical and biological environment; and
rather than advocating freedom and competition, advocates restraint
and adaptation. On the ecological perspective each species occupies
a niche in the biosphere, and human well-being can only be maintained in the long run by humankind as a whole staying within the
bounds of a sustainable niche.[18] This is not a static view however,
for, since the environment is constantly changing, so too are the
bounds of such a niche.

---

[17] In this latter regard, cf. Page (1991), esp. p. 71.
[18] It has been argued by some that competition is a key aspect of biological evolution, while this has been denied by others: for the latter view see e.g. Clark (1991),
p. 403 and Wikstrom & Alston (1992), pp. 3–4. Our concern here however is to
present the core of what we have termed the ecological view on sustainable development, and it may safely be said that on that view it is the notion of adaptation that is
emphasised.

These differences with regard to basic concepts, values and perceptions of human action mean that the economic and ecological perspectives are perspectively incompatible in each of these regards, independently of how they respectively apply to the particular issue of sustainable development, the common denominator being an atomistic/holistic difference of orientation.

Looking more particularly at their respective approaches to sustainable development, the view from the economic perspective is that through a continual increase in self-interested trade the total amount of benefits to humans can be constantly increased, and that development in this sense is a prerequisite for ensuring its own sustainability. In order actually to achieve this end however, a sufficient amount of the increase in benefits must be devoted to providing and using cleaner and less resource-demanding technology. To accommodate this requirement the environmental benefits to be gained by the use of such technology must be brought into the monetary market system (must be 'internalised'). The means of doing so thereby become a focal point of environmental concerns on the economic perspective, manifest for example in the suggested levying of pollution taxes, in the imputing of shadow prices to relatively scarce ecological benefits, and in the employment of such research techniques as cost-benefit analyses and contingent-valuation methods.

In comparison, the ecological perspective sees the problem of sustainability to concern the biosphere as a whole, and its solution to lie in decreasing the impact humans have on it, while development is to take a form not measurable in terms of economic growth. The main way by which sustainability is to be achieved is by reducing physical consumption, thereby reducing both resource-use and pollution. Herein lies the essence of the perspectival incompatibility of the two perspectives viewed as directives for attaining sustainable development, the economic perspective advocating increasing production which goes hand in hand with increasing consumption, while the ecological perspective advocates decreasing consumption which implies decreasing production. Here it might be countered that increasing economic activity need not imply increasing physical production. But the economic perspective does not recognise this distinction in its advocacy of attaining sustainability through economic growth, as is evidenced by the views cited near the beginning of this appendix.

How decreasing consumption is to be made compatible with some form of development thus becomes a central issue on the ecological approach. In this regard Herman Daly, for example, has suggested replacing the notion of quantitative economic growth with qualitative economic development.[19] Exactly what form the 'development' of 'sustainable development' should take on the ecological perspective is still open to discussion however, including whether it should be conceived as economic in any sense at all. Taking 'development' in a broader sense, the consensus would seem to be that it should include making possible the realisation of people's various talents and abilities, as well as the fulfilment of their need to feel that they are part of a community, in such a way as is in harmony with the environment.[20]

Thinking along these lines brings out what may be a rather fundamental difference in the way development can be conceived according to the two perspectives. Rather abstractly, we can imagine two kinds of development, one being progress relative to a starting-point, and the other, relative to a goal; and we suggest that 'economic' development (growth) is of the former sort, while 'ecological' development may be of the latter. Though these two notions need not exclude one another, they do so in application to the present case in the way to be suggested here. Thus where 'economic' development consists in increasing total benefits to humans, 'ecological' development may be conceived as consisting essentially in nothing other than approaching the goal of sustainability. An apparent advantage with

---

[19] Cf. Daly (1987), p. 224: "By 'growth' I mean quantitative increase in the scale of the physical dimensions of the economy; i.e., the rate of flow of matter and energy through the economy (from the environment as raw material and back to the environment as waste), and the stock of human bodies and artifacts. By 'development' I mean the qualitative improvement in the structure, design, and composition of physical stocks and flows that result from greater knowledge, both of technique *and of purpose*."

[20] In this regard cf. e.g. Meadows et al. (1972), p. 175; and Clark (1991), p. 413: "A growing group of thinkers from various disciplines is converging on the same overall vision: a globe with thousands of locally managed, self-reliant economies, based on ecologically meaningful boundaries and comprising culturally and historically integrated communities. The goal is to strengthen meaningful participation in a shared community while creating identification with the communally managed local resource base."

the 'economic' notion is that it is in principle applicable given any state of the world – no matter how rich we all are we can always imagine ourselves becoming richer – whereas on the 'ecological' notion, once sustainability is attained, it would seem that development would come to an end. In practice, however, as intimated above, the opposite may be the case. Given the earth's finite resources, it can be argued that there is a limit to how rich we can become; and given that reaching sustainability is a matter of adapting to an ever-changing environment, it can remain a goal constantly to be pursued as the environment changes.[21] What is attractive about this latter conception of development for the ecological perspective is that it suggests that self-realisation consists essentially in living in harmony with one's environment, while at the same time allowing that such self-realisation need not be restricted to humans.

Due to their different ways of conceiving of and valuing human action and what it can affect, the economic and environmental perspectives are perspectivally incompatible. As mentioned above, it is to be noted that the perspectival incompatibility of the two views is not dependent on their each suggesting that human activity be different from what the other suggests it to be. As in the case of the Necker Cube, where a drawing of a cube can be seen as showing it from above or below, the way incompatible perspectives actually manifest themselves may in fact be the same. In the present case this means that though the economic and ecological perspectives involve disparate ways of conceiving of and valuing human action, in application they may nevertheless suggest that the same actions be taken in order to achieve their respective aims. And though their aims may be different, they might be fulfilled by the same state of affairs, as is implicit in considering both perspectives to have sustainable development as their goal. In order that one of them might be considered superior to the other however, they must differ in the actions they respectively advocate for achieving that goal.

---

[21] As is in keeping with a view expressed by Clark (ibid., p. 411): "'[S]ustainability' cannot be regarded as the maintenance of a fixed set of ecological parameters. Policy makers, academic ecologists, and indigenous peoples alike will be faced with a moving target, whose direction of movement may not be readily predictable."

4. EMPIRICAL CATEGORIES AND OPERATIONS

The empirical categories of a perspective are those in terms of which it makes contact with empirical reality, and the relevant operations are those involved in the actual application of the perspective to that reality. In the case of perspectivally incompatible perspectives, at least some of the categories they involve must be the same, a state of affairs which is evidenced by the perspectives' employing the same or similar operations in their application to empirical reality.

In the present case then we must ask what the categories are in terms of which the economic and ecological perspectives on sustainable development make contact with empirical reality, and whether they are the same. In the present regard we might say that both perspectives share the basic categories of 'sustainability' and 'development.' In other words, with regard to how well the respective perspectives apply to the empirical world we must consider them in terms of sustainability and development conceived as empirical phenomena. Each of these categories however constitutes a blanket-concept, under which are more specific concepts which constitute the criteria with regard to which the applicability of the blanket-concepts and ultimately the perspectives themselves is to be judged. As regards 'sustainability' such criterial concepts include 'pollution,' 'resource depletion' and 'biological diversity'; and as regards 'development' they might include 'real income,' 'self-realisation' and 'disalienation.'

As regards 'development' in particular we see that, unlike in the paradigmatic scientific case, the actual criteria in terms of which the perspectives are to be judged to be applicable may be conceived differently within each perspective, which increases the difficulty in determining their relative acceptability. If the economic perspective aims to increase real income while the ecological perspective aims to facilitate a sense of community, i.e. if they are out to accomplish different things, then it can be difficult to compare them as regards how successful they are or might be.

The second question concerns the *operations* to be employed in applying the categories – or more particularly the criterial concepts that fall under them. As regards sustainability, the applicability of the criterial concepts of pollution, resource depletion and so on are common to both perspectives, and can in fact be measured by operations relevant to both of them. But the problem in the present case, which can also arise in certain scientific cases, is that it is exceedingly difficult to determine what the results of the relevant operations should be according to the respective perspectives. As will be seen below however, it may be possible to circumvent this difficulty, as well as that regarding the potentially diverging aims of the two perspectives with regard to development.

5. RELATIVE ACCEPTABILITY

The relative acceptability of competing perspectives is to be judged in terms of three basic criteria: their relative accuracy, scope and simplicity. In the scientific case these criteria are to be thought of in terms of the explanatory function of scientific theories. Thus the explanation provided by one scientific perspective might be more exact than another, in which case it would be considered more accurate; or it might be able to explain more sorts of phenomena, in which case its scope would be greater;[22] or the explanation it provides might involve fewer ad hoc modifications, in which case it would be simpler.

In the present case the question concerns which of the economic and ecological perspectives, assuming alternately that each of them has obtained a widespread following, is the more likely to lead to the global attaining of sustainable development. In this context the notion of accuracy does not seem all that relevant as a criterion, for what is important is that the general aim of sustainable development be reached, and not that it take a previously specifiable form. It may be thought that the economic perspective is more accurate than the ecological due to its concern for well-defined goals and its emphasis on quantitative research methods. But the clear definition of the tools used and aims pursued does not ensure that the aims will be met, which is the question at issue here. As regards development it is to

---

[22] Our three criteria bear some similarity to the three criteria Ward suggests for comparing alternative economic worldviews. His second criterion corresponds to ours of accuracy and scope. With regard to it he says: "The question is, Do the world view's assertions fit the known facts? In practice, this criterion turns out to be rather less specific than it sounds because of the nature of the 'facts' that appear as parts of so broad a doctrine as a world view." (1979, p. 270).

be noted that the form suggested by the economic perspective might well differ from that of the ecological perspective; but given this it makes little difference whether either of the perspectives can determine in advance specific details regarding the form it is striving to attain.

The potential scope of both perspectives may be assumed to be the same, namely human activity generally – though in this regard it might at least be mentioned that the economic view has a particular focus on economic activity between humans, while the ecological view considers the environment as a whole, of which inter-human economic activity is only a part. But this difference does not mean that the economic view need have a less widespread effect, which is what the notion of scope should concern here.

The notion of *simplicity* is perhaps of greatest interest in the present case, for it concerns the internal coherence of a perspective,[23] and it is just this internal coherence that has been called into question with respect to the economic view. In this regard Inge Røpke, for example, raises the following point:

[T]hat people will be willing to devote more resources to environmental ends when their incomes grow does not necessarily counterbalance the fact that they will be consuming and polluting more when they become richer. Similarly, the development of cleaner technologies as a part of the growth process, resulting in less pollution and less use of resources per unit of production, does not guarantee that the effects of growth will be sufficiently counteracted.[24]

And a number of contributors to the issue have claimed that the idea of sustainable development, if development is to be understood as unqualified economic growth, is actually a contradiction in terms. Frank B. Golley, for example, notes that:

---

[23] In this regard cf. Ward's third criterion: "The remaining test has to do with whether or not the various parts of the world view fit together to make a coherent whole. This is not a simple criterion to apply because there is a certain open-endedness to it; for example, a revealed incompatibility may be relatively minor, or one might feel that it can probably be corrected without too much difficulty. But major incompatibilities are serious obstacles to the acceptability of a world view." (ibid.).

[24] Røpke (1992), p. 7. See also Schumacher (1973), p. 127, and Gahrton (1988), p. 21. Next quote, Golley (1990), p. 16; cf. also Porritt (1992).

If development is defined as expansion of the physical structure of the built environment or numbers of people, sustainable development is an oxymoron. Development can never be sustained indefinitely since it requires resources for expansion and an environment to receive wastes, and neither resources nor environments are infinite.

The only way the economic perspective might circumvent this fundamental inconsistency would be to distinguish between economic growth that implies physical growth or increase in throughput of matter/energy, and economic growth that does not, and then go on to advocate only the latter. Apart from the fact that such an ad hoc move is not being recommended by any advocates of the economic view, however, it would undermine the very reasoning on which that view is based, according to which an improvement in material well-being is to provide the means for solving the world's environmental problems.[25]

The essential incoherence of the economic perspective pointed to here has been made perspicuous in a theory advanced by Richard G. Wilkinson. According to Wilkinson, the very idea of property – fundamental to the economic view – indicates a need to make a claim to limited resources, and economic growth is essentially a re-action to the scarcity engendered by an environmentally unsustainable situation. On his theory, which is partly Malthusian in thrust, when a human population has lost the ability to limit its size, its growth leads to greater demands on the environment such that the only way that the increased population level can be maintained is by discovering ways to exploit the environment more thoroughly. Thus: "Within a stable society in ecological equilibrium, population growth is the most dangerous threat to continued stability."[26] Under such conditions necessity can well become the mother of invention, with new technologies being developed, the implementation of which gives rise to increased economic activity. In Wilkinson's words:

Once one has the concept of a society existing in ecological equilibrium there is no difficulty in accepting that the development of need is the real cause of economic development. ... Development is primarily a matter of

---

[25] Cf. Daly (1992), p. 126: "But as the growing weight of anomaly complicates thinking within the growth paradigm to an intolerable degree, the steady-state view will become more and more appealing in its basic simplicity."

[26] Wilkinson (1973) p. 57; next quote, pp. 63, 99, 105.

increasing the rate of environmental exploitation to support a growing population. ... Instead of regarding development as a matter of 'progress' towards a 'better life' motivated by an incurable dissatisfaction with our present lot, we see that it is a process of solving a succession of problems which from time to time threaten the productive system and the sufficiency of our subsistence. In effect, human societies out of ecological equilibrium have to run to keep up; their development does not necessarily imply any longterm improvement in the quality of human life.

A paradigmatic instance of this is the industrial revolution in England, where population growth led to a shortage primarily of wood, the (inferior) substitute for which was coal. This set the stage for such technological advances as the invention of the steam engine, needed to pump water out of the ever-deepening coal mines. In this context it is to be noted that, as compared to the task of harvesting wood in a sustainable manner, that of mining coal is much greater, and increases as mines are dug deeper; also, the task of transporting the fuel is greater, due to the availability of coal only in particular locations. While the environmental results of the transition to coal as the primary fuel (made possible through technological innovation) were the depletion of a non-renewable resource, increased pollution and the establishment of a huge transport network of canals and rails, the economic results were 'growth' in the form of an increase in per capita and thus overall consumption of energy and materials in order to meet daily needs:

[I]ndustrialization requires a more extravagant lifestyle than the modes of production that preceded it. The problems it creates and the needs it sets up make increased consumption a necessity if people are to lead reasonably satisfactory lives. The continuous expansion of gross national product which this requires should perhaps be regarded more as a reflection of the rising real cost of living than an indication of increasing welfare.[27]

Thus, Wilkinson reasons, as regards our present predicament, "there must be serious doubts about the wisdom of continued economic growth *per se*. Present and future ecological problems and the demands of social welfare make it clear that we should be much more discriminating about the forms which development takes."

---

[27] Ibid., p. 185; next quote, p. 194.

Here we may supplement Wilkinson's account by noting that some of the growth resulting from the industrial revolution was manifest in the steam engine's being put to other uses, and in the exchange of excess coal for other commodities. More generally, we see that once a new line of technological development has been opened it tends to be pursued,[28] and once a resource has been tapped it tends to be exploited, thereby providing increased material benefits that 'overshoot' the requirements of the population in question.[29] In this way technological innovation can have the effect of suddenly increasing the potential for a particular area of land to support human habitation, thereby constituting a major factor in the population's losing its ability to control its own numbers. And as the population grows relatively unchecked, new technology becomes needed to support it when it becomes too large. Thus a vicious circle is created, in which economic growth is made possible only by further technological advance which in turn further degrades the environment while promoting the size of the population.[30]

Considering the current state of the world, we might say that the circle is presently making a particularly large loop, due to the accessibility of a huge though finite quantity of fossil fuels, the use of which supports a tremendous increase in the world's population at the same time as it gives rise to environmental damage on a global scale. And when the present circle closes – when the availability of fossil fuels trails off – the world will face a situation that can be nothing less than catastrophic. This circle must be broken.

The above considerations thus suggest that due to its essential incoherence the economic perspective on sustainable development is not a viable alternative in the present debate. Rather than accelerate economic growth, we should curtail it, as is advocated on the ecological perspective. On the basis of the reasoning presented

---

[28] In this regard, cf. von Wright (1986, p. 142): "[T]echnical innovations are marketed so that they meet desires which are awakened in consumers and formed by already existing technology." von Wright goes on to explain how such desires can become 'needs' once a society is accustomed to their being fulfilled, as is presently the case with regard to e.g. access to the Internet and the use of cellular phones.

[29] A view similar to that advanced by Nathan Rosenberg, cited in Wilkinson (1973), p. 146.

[30] This idea is further developed in App. I of Dilworth (2007), and in much greater detail in Dilworth (2008).

above, it may be argued that the plateau or decline in the standard of living in the West since the 1960s – despite a quadrupling of global economic production during this period[31] – is in fact itself largely the *result* of the relatively unrestrained pursuit of economic growth through technological development. Present world-wide economic problems are but symptoms of a disease consisting in our having over-exploited the environment on which our lives depend; and continuing attempts to alleviate these symptoms through a policy which essentially promotes the disease can only have one result: a successive worsening of the symptoms and the eventual death of the patient. This does not mean that the ecological perspective is itself without problems, but as regards sustainable development it is right in its view that the attainment of ecological sustainability should be a goal in itself, not one linked to the maintenance of economic growth. Furthermore, the view that development might consist in the very approaching of sustainability, as has been advanced here, adds coherence to the ecological perspective on sustainable development as a whole. Whatever notion of development one employs, however, the ecological perspective must be viewed as more acceptable than the economic perspective as a means of attaining sustainable development.

---

[31] Niles (1999).

MODERN SCIENCE AND THE DISTINCTION
BETWEEN PRIMARY AND SECONDARY QUALITIES

The exact natural sciences, of which physics is the paradigm, are often referred to as the *empirical* sciences, in contrast to the formal sciences of mathematics and logic. When thought of in this way, what is considered characteristic of them is that, unlike mathematics and logic, the knowledge they afford is obtained via *experience*.

When logical empiricism was at its height, empirical scientific knowledge was thought to be obtained by *direct* experience,[1] the idea being that the more direct the experience, the less the chance of being mistaken. In more recent years this view has received much criticism, and today there is an almost universal consensus that even the most basic of scientific observations involve an element of 'theory.' However, the efforts made in this direction have been marred by, among other things, an overlooking of the difference between theories, on the one hand, and laws and principles, on the other. Simply to deny that the empirical aspect of science consists of something like sense-data does not of itself imply that it is impregnated by theory, nor that it is not objective.[2]

The sense in which the empirical sciences are empirical can be clarified beyond the expositions given in Appendices I and IV above through a consideration of the distinction between primary and

---

[1] Thus, for example, Hempel speaks of "the insistence of logical empiricism that all significant scientific statements must have experimental import, [and] that the latter consists in testability by suitable data of direct observation" (1965, p. 125); and Ayer suggests that "It is a characteristic of ... 'basic propositions,' that they refer solely to the content of a single experience, and what may be said to verify them conclusively is the occurrence of the experience to which they uniquely refer" (1936, p. 10). A similar view can be found e.g. in Quine (1953), p. 17. What unites these authors here is the belief that the data of science, if not its very subject-matter, consists in the subjective experiences or observations of individual people.

[2] In this regard, cf. *this volume*, pp. 138–140; and Dilworth (2007), pp. 93–94 and 119.

secondary qualities, which we shall see to be essentially the same distinction as that drawn in Appendix IV between two senses of the term "empirical."[3] The primary/secondary quality distinction can in essence be traced back to Parmenides. With him, it takes the form of two Ways of enquiry, that of Truth and that of Seeming. Pursuing the former, one is guided by reason, and comes to know what is; following the latter, one is guided by the *senses*, the objects of which are but "mere names."

The first major step in the evolution of the distinction occurs with the ancient atomists, and in Democritus it takes the form of a difference between 'legitimate' and 'bastard' cognition. Democritus says: "There are two forms of cognition, one legitimate, one bastard. To the bastard belong all these: sight, hearing, smell, taste, touch. The other is legitimate, and separate from these."[4] And: "By the senses we in truth know nothing sure, but only something that changes according to the disposition of the body and of the things that enter into it or resist it."

While 'bastard' cognition is provided by the senses and does not go beyond what we directly perceive, 'legitimate' cognition is apparently obtained by something akin to what we would today call scientific theorising, and concerns the properties of the smallest constituents of matter, viz. the atoms. Thus Aristotle's pupil Theophrastus says:

Democritus does not give the same account of all the objects of sense, but distinguishes between some on the basis of size, some on the basis of shape, and others on the basis of arrangement and position. ...

Of the other objects of sense he says that they have no existence in nature, but that all are effects of our sense organs as they undergo the alteration which brings into being what appears to us. For neither hot nor cold has any reality, but the shape, "undergoing a change," works a change in us also. ...

An indication that the aforementioned qualities do not exist in nature is that things do not appear the same to all living creatures, but what is sweet to us is bitter to others, and to still others sour or pungent or astringent; and so on with the rest.[5]

---

[3] Cf. *this volume*, pp. 183–184.
[4] Fr. 11, as translated in Guthrie (1965), p. 459; next quote, Fr. 9, as translated in Burnet (1914), p. 197.
[5] Excerpted from Theophrastus' *De Sensu*, 60–64, as translated in Robinson (1968), pp. 199–200.

We are here reminded of the use of standards in measurement, as taken up in Appendix IV. By focusing on an intersubjective standard, one circumvents the problem of subjectivity. A measuring device reacts in one and only one way in a given situation, and that way can be grasped by any number of observers. Democritus never made the move to measurement to solve the problem of how to obtain knowledge of the world given the fallibility of the senses, however; and it is not altogether clear whether or how he thought that 'legitimate' cognition was to provide knowledge of the size, shape, arrangement and position of things. There is no indication, for example, that he thought such properties to be quantifiable, while denying this status to the objects of 'bastard' cognition.

Taking Democritus' notions of size, shape, position and arrangement to represent primary qualities, we may call his general conception of such qualities an *a priori* one: as conceived by Democritus, nothing can be a body without having these properties. This 'a priori' conception of primary qualities is also that of Lucretius,[6] and is much later taken over by Galileo, though with Galileo we see the beginning of a movement to a distinct though overlapping notion. The continuity of thought from Democritus to Galileo is, however, clear in the latter's *Assayer*. There Galileo says:

I feel myself impelled by the necessity, as soon as I conceive a piece of matter or corporeal substance, of conceiving that in its own nature it is bounded and figured in such and such a figure, that in relation to others it is large or small, that it is in this or that place, in this or that time, that it is in motion or remains at rest, that it touches or does not touch another body, that it is single, few or many; in short by no imagination can a body be separated from such conditions: but that it must be white or red, bitter or sweet, sounding or mute, of a pleasant or unpleasant odour, I do not perceive my mind forced to acknowledge it necessarily accompanied by such conditions; so if the senses were not the escorts, perhaps the reason or the imagination by itself would never have arrived at them. Hence I think that these tastes, odours, colours, etc., on the side of the object in which they seem to exist, are nothing else than mere names, and hold their residence solely in the sensitive body; so that if the animal were removed, every such quality would be abolished and annihilated.[7]

---

[6] Cf. Lucretius, Bk. II.

[7] From *Il Saggiatore* (*The Assayer*), in *Opere Complete di Galileo Galilei* (Firenze, 1842), IV, p. 333, as translated in Burtt (1924), p. 85.

Galileo, like Democritus, is concerned to suggest that what we might call secondary qualities do not exist in nature independently of us; and, like Parmenides, he speaks of the objects of sense as "mere names."

In further characterising secondary qualities as conceived by Democritus and Galileo, we might say that, due to the way that they reside in the perceiving subject, they are never determinable by more than one sense. We cannot see tastes, smell sounds or hear colours. The primary qualities, on the other hand, are not tied in this way to particular senses. One can, for example, often determine the size or the position of a thing by either sight or touch. This further supports the idea that primary qualities are not resident in the sensing individual, and is of particular relevance to measurement. Though measurement is normally performed with the aid of sight, touch might often replace sight in practice, and is perhaps always able to replace it in principle.

The distinction between primary and secondary qualities was taken over not only by Galileo, but in various forms by virtually all of the scientifically most influential thinkers of the seventeenth and eighteenth centuries, including Bacon, Kepler, Descartes, Hobbes, Boyle, Newton and Locke.[8] While for the most part these individuals retained the 'a priori' conception of primary qualities, as has been argued by E. A. Burtt there was at the same time a tendency to conceive of such properties as those which are susceptible of *mathematical* treatment.[9]

Burtt does not explain this correlation, and a step towards doing so might be to note that primary qualities occupy a persisting world common to us all, while secondary qualities exist in a subjective world, which is constantly in flux. This means not only that primary qualities tend to persist in things while secondary qualities come and go, but that by being properties of external objects primary qualities can be compared in a way which makes little sense in the case of secondary qualities. One can compare the length of one physical object with that of another, but how should one compare two experiences of

---

[8] Concerning Locke, see his (1690), esp. Bk. II, Ch. VIII; for Bacon, see e.g. Cranston (1967), p. 239; as regards the others, see e.g. Burtt (1924), pp. 66–68, 117–121, 130–132, 181–182 and 235–237.

[9] In this regard, cf. ibid., pp. 67–68, 83–84, 117–118 and 306.

the colour red? Mathematics enters in the case of length, it would seem, when we have two objects of the *same* length which we consider joining end to end, that is, the lengths of which we consider *adding*. More generally, it is the *additivity* of certain of the primary qualities that makes them open to mathematical treatment.

But what, in turn, is it to say that an empirical property is susceptible of mathematical treatment? In a very straightforward way it is simply to say that it is measurable.[10] And in this way then we can move to aligning primary qualities with measurable properties.[11]

The subject-matter of science is not reality as it appears, but as it actually is. While empirical science would be impossible without subjective experience, the objective results it obtains are the results of attempts made, through measurement, to acquire knowledge that goes *beyond* such experience. In this regard it has been successful, and has provided us with knowledge not of our experiences, but of reality itself.

---

[10] On this point, and that of the previous paragraph, see further Campbell (1921), pp. 117f.

[11] For other discussions of the identification of primary qualities and measurable properties, see Campbell (1938), pp. 128ff., and Hirst (1967).

A THEORY OF IDENTITY AND REFERENCE

The provision of a philosophical theory – in the present case a theory of identity and reference – differs in kind from the provision of a scientific theory. Where a scientific theory is intended to explain empirical (mensural) phenomena through indicating their cause or causes, a philosophical theory is intended to make sense of a particular subject-matter by providing a coherent way of thinking about it. The requirements that we place on such a theory are that it be consistent, that it be sufficiently broad to apply to the majority of interesting cases, that it be coherent – i.e. that it have a kind of unity and not require ad hoc additions – and that it be reasonable, in some generally accepted sense of this term.

1. TO ANALYSE A PARTICULAR PHENOMENON

There are a variety of ways of going about creating a philosophical theory, involving various sorts of presuppositions. Such presuppositions concern, for example, the nature of the subject-matter to which the theory is being applied, and what is to count as reasonable in dealing with that subject-matter. The tack taken here in providing a theory of identity and reference – which largely parallels the presentation of the theory of scientific progress in the main text – is to begin by analysing a particular empirical phenomenon in which concepts at least similar to those that may be used in the theory can be found. Then, the relations among these concepts in the phenomenon having been indicated, the concepts are abstracted from it, and a coherent way of thinking – i.e. a theory of identity and reference – is created using them. Following this, the theory is applied to some of the better-known instances of the problems of identity and/or reference as they appear in the literature. And this presentation is then rounded off with a critique of the alternative theory that gave rise to these problems in the first place.

From the above we see then that the theory to be presented here itself consists in a conceptual scheme based on concepts obtained from the phenomenon, but which have been raised to a more abstract level and universalised. In this way the phenomenon is the *source* of the theory; but it may also function as an *analogue* in *applying* the theory, in which case its combined roles make it a *source-analogue*.[1] And the theory itself, in its application to particular problems of identity and/or reference, should in the best of cases be able to solve them through showing how their constituents can be conceived in such a way that the problems do not arise. Thus in order for this approach to be successful it is necessary that the example chosen be sufficiently complex to allow the development of an abstract conceptual scheme capable of handling a wide variety of cases.

2. THE GESTALT MODEL

What is here chosen as the starting point for our theory is a particular gestalt-switch phenomenon, namely Hanson's bird-antelope (presented on page 56n. of this book), which is a particularly well-suited such phenomenon for a reason that will become apparent at the end of this section. Thus this phenomenon will function here as the *Gestalt Model*, providing an intuitive and easily understandable example that can serve as the starting point for the construction of our theory. As in the case of the Gestalt Model as used in the main text, however, and as intimated above, once the theory is constructed the model is to function solely as an aid to understanding the theory.

Objects, Things and Identity

Taking Hanson's bird-antelope as the Gestalt Model then, we note first the distinctiveness of 'level' between, on the one hand, seeing the bird-antelope as a drawing of a bird or an antelope, and, on the other, seeing it as a mark on a piece of paper. What we might say is that what it *is* is a mark on a piece of paper, which can be *seen* as a drawing of either a bird or an antelope. Taking *things* to be what

---

[1] As regards sources, analogues and source-analogues, see e.g. *this volume*, pp. 187–189. In the present case the phenomenon may also be seen as a *paradigm* for the application of the theory.

exist, and *objects* to be objects of thought, we might call the mark of which both the bird and antelope aspects of the figure are manifestations a thing as relative to these aspects, each of which may be called an object. This distinction is a relative one, however, and we could, for example, consider the seeing of the bird-antelope simply as a mark also to be an aspect of it, i.e. to be an object, albeit one on a different level than the bird and antelope aspects. In this case, then, the mark would be both an object and a thing. Similarly, we can imagine a situation in which the entity is e.g. a real antelope, in which case the antelope aspect would not only be an object but also a thing. In both of these latter cases the object and the thing would be identical.

Intention, Reference and Referent

When looking at the bird-antelope as, say, a bird, we can, following the development of the Gestalt Model in the main text, consider ourselves to be applying a concept to the figure. This application consists in giving an *intention* to the concept. When a person looks at the bird-antelope and sees it as a bird, he (or she) is *'intending'* the concept 'bird' to the figure, or *applying* the concept to the figure. In so doing, one provides the concept with a *unique* referent, viz., the figure conceived of as a drawing of a bird.

This 'intention' is not inherent in the concept itself, for the concept 'bird' can be applied to things other than the bird-antelope, and to things which closer inspection shows not to be birds at all – as would be the case were the bird-antelope to be a real antelope, as above. In fact it may even be applied to nothing at all: the concept can be applied to a mirage, or to the effects of an hallucination. All that is required is that the person intending the concept *conceive* of there being an object. Thus, the intention derives from the person doing the intending, and, within certain limits, he has the freedom to intend (conceive of) what he will as a bird or other entity.

The notion of intention being employed here is the same as that in the main text, and is closely aligned with that of reference. But, as regards reference, where intention is originally a *pre*-linguistic act involved in conceiving of something as something, reference is here being conceived of as a *linguistic* act[2] in which the person referring not only is himself applying a concept to a particular thing, but *through the use of a term or terms* (in speaking or writing) is attempting to get another person to do so as well. Thus when one refers to the bird-antelope as a bird, at least part of what one is doing is trying to draw some other person's attention to the bird aspect through the use of words. One attempts linguistically to get that person to pick out the unique object to which it is intended that the concept be applied; and the success of the reference depends on whether one has done so. When referring to the bird-antelope as a bird, the bird aspect of the figure is the direct *referent* or *object* of the act of referring; and we can consider the act itself to be a *reference* to the bird aspect. Thus we are here making a clear distinction between a reference, which is an act, and a referent, which is an object.

Reference can also be successfully made, however, even if the listener should be unable to pick out the referent. For example, I might refer to the bird-antelope in a conversation with a person who is unfamiliar with it, but is familiar, say, with the duck-rabbit. Upon being asked by him what the bird-antelope is, I can reply, "it's a gestalt-switch figure, like the duck-rabbit," and though my saying so does not provide him with enough information to pick it out, my reference may nevertheless have succeeded to the extent necessary in that situation.

Sense and Meaning

While the term "bird-antelope" may be used in referring to a wide variety of objects, under normal circumstances its use is constrained by its *meaning* or *sense*. It is this meaning or sense that leads to the term's normally being used in referring precisely to instances of the bird-antelope. And the same may be said of the phrases "bird aspect" and "antelope aspect." The senses of the two terms differ such that the former is normally used in referring to the bird aspect of the bird-antelope while the latter is normally used in referring to the antelope aspect.

---

[2] Thus this sense of "reference" differs from that of the main text; cf. *this volume*, pp. 61&n., 80n. and 95.

This sense or meaning is nothing other than the primary *concept*[3] one applies in using the term. Note however that not only does a term's sense not *dictate* its use[4] – entities other than the bird-antelope *can* be referred to as the bird-antelope – but it allows of its being used to refer to various entities in various contexts. Thus, for example, I can use the term in referring to the bird-antelope as it is printed in this book, or in Hanson's.

Referring vs. Naming vs. Describing

It may be noted that when referring to the antelope aspect of the bird-antelope, saying e.g. that it is looking to the right, one is not naming that aspect. If one wants, one can give it a name – let us call it "Andrew." This is naming ('baptising') the antelope aspect, which does *not* consist in giving an ostensive definition of the term "Andrew." Apart from the fact that names are not susceptible of definition, it is clear in the present case that no definition is intended. As regards naming, we note further that if on some future occasion we were to speak of the antelope aspect as Andrew, we would not be naming it then, either. Rather, we would be referring to it by name, or using its name in referring to it. And where the term "Andrew" may connote a male entity, we could just as well name the antelope aspect "X," a term which (at least ideally) should have no sense or connotation at all.

Here again it is a matter of a *person* doing the referring.[5] Andrew's name does not pick him out; we do, either pre-linguistically or through being referred to him by someone else who uses his name or other relevant words. Andrew can be linguistically picked out in different ways, the success of which often depends on the context in which we find ourselves. If it turned out that the bird aspect of the figure was also called "Andrew" – and nothing prevents this from

---

[3] Cf. Church (1951), p. 196.

[4] As many philosophers would take it to do. In this regard cf. e.g. Feyerabend's 'epistemological realism' (*this volume*, pp. 78–79).

[5] Cf. Strawson (1950), p. 111: "'Mentioning,' or 'referring,' is not something an expression does; it is something that some one can use an expression to do;" and p. 113: "it is people who mean, not expressions." Here, however, we should replace "expression" with "term."

being so – then someone's simply using the name "Andrew" would not suffice for others to delimit the antelope aspect.

Also, it is important to distinguish between referring to one of the aspects of the figure (or to the figure itself), and describing it. One may of course describe an aspect in referring to it, in order to help another pick it out; but in such a case what is of interest is whether the description functions to that end, and not whether it is perfectly true of the aspect in question, or even whether the aspect actually 'falls under the concept'[6] intended in referring to it. Thus we might want to draw another person's attention to the bird aspect by describing it as a drawing of a pelican whose head is pointed to the left, and which has a gaping mouth, and so on; but, quite possibly, assuming that the person who drew the figure could determine what the figure is of, that aspect may not at all be a drawing of a pelican, or even of a bird (it might be e.g. a pterodactyl) for that matter. Similarly we might want only to draw attention to the figure, and not to it under a particular aspect; and this too we could perhaps do by referring to it as a drawing of a pelican. Thus an object may be successfully referred to despite being misconceived by both the person referring and his listener. But as long as the description suffices for the listener to delimit that particular aspect, whether or not *as* that particular aspect, it may have served its purpose, depending on the context.

Identity and the Analytic vs. the Synthetic

Is the bird aspect identical to, or the same as, the antelope aspect? No, they are *different* aspects of the *same* figure. On the Gestalt Model they are different objects each of which is a particular manifestation of one and the same thing. Using the notion of identity in this context, we should say that the thing of which the one object is a manifestation is *the same as* or *identical to* the thing of which the other is a manifestation.

To say of the bird-antelope that it is identical with itself is *tautological* or *analytic*. On the other hand, to say that Andrew is the same as the antelope aspect of the figure may or may not be analytic, depending on the situation in which we say it. For example, the person we are speaking to may have heard us use the name "Andrew"

---

[6] Cf. *this volume*, p. 63n.

before, and wonder who or what Andrew is; or, on the other hand, he may be being informed that the antelope aspect is hereafter to go under the name "Andrew." In this latter case then, to say that Andrew and the antelope aspect are the same might be considered analytic, while in the former it might be considered synthetic. Similarly, if the antelope aspect should have *two* names, say, "Andrew" and "Anthony," then to say that Andrew is Anthony may or may not be analytic, depending on the context. Or, if we were to use the notion of identity with regard to the Gestalt Model as a whole, our saying for example that the bird-antelope and the gestalt-switch diagram on page 56n. are the same thing may be either analytic or synthetic, depending on the context.

Equivalence and Equality

In considering notions similar to identity, it may be pointed out that we should not say that the antelope aspect of the bird-antelope is *equivalent* to the bird aspect. Using the notion of equivalence in the context of the Gestalt Model, we might say rather that at least in certain situations the *duck-rabbit* is (potentially) equivalent to the *bird-antelope* when it comes to functioning as the model.

As regards equality, if we consider the form of the bird-antelope here being used as the Gestalt Model, viz. that in which it contains five wiggly lines, and take the lines to represent feathers in the bird aspect and marks on the pelt in the antelope aspect, we could say that the number of feathers is *equal* to the number of marks, or that the number of feathers *is the same as* the number of marks. This, however, is of course not to say that the feathers are identical with the marks. Apart from their number, what is the same in the present context are the things (lines on a piece of paper) of which they are manifestations.

3. A THEORY OF IDENTITY AND REFERENCE

On the basis of our presentation of the Gestalt Model we are now in the position to construct a theory of identity and reference. Should questions arise as to what is intended in the theory, they, or at least most of them, should be answerable by considering the model. As mentioned, however, our theory, like the Perspectivist conception of science, should be able to stand on its own, and reference be made to the Gestalt Model only for purposes of clarification.

Objects vs. Things

According to our theory then, reference is a *linguistic act* and is to be distinguished from what is being referred to, the *referent* of the act, which is an *object*. The act of reference is performed by a *person* – the *speaker* (or writer) with the purpose of drawing another person's, the *listener's* (or reader's), *attention* to the object. This act of reference involves the use of *terms*, which brings a *concept* of the object to the listener's mind. That an object is an object depends on its being conceived as such, and thus we may say that objects of reference, and objects of intention more widely, are *intentional objects*.[7] Intentional objects need not exist other than as intended by the speaker and/or listener, though they *may* also exist as a matter of fact.

Things, on the other hand, do exist as a matter of fact; and they may, or may not, be objects. Normally they lie at a deeper epistemological level than objects, a level which, in certain cases, may be quite inaccessible. Another way of expressing this is to say that relative to the *ontological* level of things *qua* things, objects lie on an *epistemological* level, the idea being that what we know or are acquainted with are the aspects of things, while things themselves are what really exist. This distinction bears obvious relation to that between primary and secondary qualities (as in Democritus), the nominal and the real (as in Locke), and the phenomenal and the noumenal (as in Kant). With regard to primary and secondary qualities we might consider primary qualities to be properties of things, and secondary qualities to be properties of objects. And as regards the nominal and the real we might say, for example, that it is only through the way that a (real) thing is manifest as a (nominal) object that it can be *identified*.[8] Similarly, with respect to the phenomenal/noumenal distinction, we could say that the deepest ontological level (the noumenal) must be unknown to us, for we can only have concepts of objects (phenomena).

---

[7] For the use of this notion in another context, see Dilworth (2007), p. 249.
[8] Cf. the discussion of the relevance of difference of level in identifying *natural kinds* in ibid., Ch. 7.

Intention, Reference and Referent

The directedness involved in reference, such that the concept is applied to one particular thing rather than another, is the *intention* given to the concept. Through giving intention to a concept one *intends* an object (i.e. a thing – or nothing – conceived of in a certain way). Thus to *direct one's attention* to something implies *intending* an object, or *performing an objectification*. And the original such objectification is pre-linguistic. In referring, what the intention given a concept does, or what applying a concept does, is provide it with a *unique* referent (this is something a definition does not do). What is to count as the object of a particular act of intention is not dependent on the nature of the concept being applied, but on the act of intention itself. To be an object of reference, what is being referred to need only be *taken* as existing,[9] and need not even be the manifestation of a thing. Thus we can ask whether such an object, or, more generally, any particular instance of an intentional object, is taken as existing as such by more than one person, or whether it exists independently of being conceived, i.e. as a thing.

All the same, we can still speak of referring to a thing, it being understood however that the thing in question is being conceived under a particular aspect – an aspect which may be identical with the thing itself *qua* thing. Any act of reference involves the application of a concept. And a thing may be referred to in different ways through applying different concepts to it, as when one refers to the bird-antelope either as a bird or as an antelope. In this case the bird or antelope aspect of the figure is (or would normally be) the *direct* object of reference, while the bird-antelope, *qua* bird-antelope, is the *indirect* object; and, should the aspect under which a thing is being referred to be identical with the thing itself, as would be the case were one to refer to the bird-antelope *as* the bird-antelope, then the thing itself would be the direct object of reference.

---

[9] Cf. Frege (1892), p. 79, for a similar conception: "'You have been speaking without hesitation of the moon as an object; but how do you know that the name [*sic*] "the moon" has in fact a referent?' I reply that it is not our *intention* to speak of the image of the moon; ... we presuppose a referent here. [I]t will suffice ... to refer to our intention in speaking and thinking in order to justify our reference to the referent of a sign; even if we have to make the proviso: if there is such a referent." The translation is partly mine; emphasis added.

In referring to a thing you *make* it an object – either direct or indirect – of reference. Conceptually, you start with (ontological) things, and then, through the act of intention, you create (epistemological) objects. And it may also be possible to say (in the case of referential phrases) that a particular objectification is more or less satisfactory, or correct or incorrect, i.e. that the concept being applied fits the object better or worse. Note however that it is not the intention that determines this, since, among other things, intentions do not characterise.

A referent is an object which may or may not be a thing, while a reference is a (linguistic) act. Of course we can refer to references, which means they can also be objects. But in one and the same act of referring the reference being made is normally distinct from the object being referred to. (Self-reference paradoxes may of course arise when this is not the case.)

Here we also distinguish between referring and referring successfully. I can refer to something without my listener having any idea what I am talking about, i.e. without his being able to form any concept of what I am referring to as being what I am referring to. Or I can make a reference thinking that what I am referring to exists either as an object or a thing when it does not. Such references would be unsuccessful however, in the first case in a way similar to my fishing trip's being unsuccessful if I come home without any fish despite there being fish in the lake, and in the second case similar to its being unsuccessful when there are *no* fish in the lake. But in both cases I have nevertheless referred despite my references' being unsuccessful, just as in both instances of my fishing analogy I have been fishing despite my not having caught any fish.

That my reference be *successful*, on the other hand, means that my audience 'gets my message' (that I catch my fish), or 'grasps' what I intend. Among other things, this presupposes that there *is* an object I am referring to; and it implies that the listener, through being able to form a concept of that object, has been able to *delimit* it – which may include his being able to *pick it out* or *identify* it, but not necessarily. If I say "my friends like me," then, under normal circumstances, I have successfully referred to my friends – my audience has been able to *delimit*, though not pick out, the objects of my reference – some certain people. But if by "my friends" I do not mean people, but *dogs*, then my reference has failed if my audience takes me to mean people: they have been unable to delimit the intended referents. Whether a person has or has not delimited a particular referent, or whether they have done so to the extent necessary for the reference to be successful, will often depend on the context, including the states of mind of the speaker and the listener. That the listener cannot *pick out* the referent of a successful reference is of course often the case, as when one refers e.g. to historical figures or non-empirical objects. It should also be noted that reference may successfully be made despite the listener's forming an incorrect concept of the object of reference. The key point is that the concept he has suffices for him to delimit the object to the extent necessary given the context. In all such cases, however, the listener must have *some* concept of the referent, which furthermore must itself exist.

Sense, Meaning and Connotation

An act of reference makes what is being referred to an object of reference and involves the application of a concept to that object – a concept which in certain cases may be more or less adequate in characterising the object. When the term used in referring is not a name, the concept may be conceived of as the term's sense, "sense" here being taken as synonymous with "meaning." And the function of a definition is precisely to indicate a word's meaning or sense (and not its referent).

The situation as regards *names* is different, however. Names have no sense, and thus do not stand for concepts which may or may not be in keeping with the relevant object(s) of reference: they do not characterise the objects they are being used to refer to. Nevertheless names *do* stand for concepts, such concepts being more particularly *notions* (in a sense similar to Kant's),[10] notions not being part of a term's meaning or sense. Thus different names stand for different notions. However, since names lack sense they cannot be defined; though, like terms with sense, they may have *connotations*.

A word's connotation consists of a non-notional concept or concepts associated with the word that, like notions, are not part of its meaning. Thus we might say, for example, that while the nouns "swine" and

---

[10] Cf. Kant (1781, 1787), A 320, B 377.

"pig" may have the same meaning, they have different connotations; that while the name "Andrew" has no sense, it does have the connotation 'male'; and that the letter "X" used as a name has neither sense nor connotation.

Further, in keeping with Strawson, we should say that the relation of terms to their senses is normally relatively constant, unlike their relations to what they are used in referring to, since terms with senses can be used to refer to various objects in various contexts.[11] The definition of a referential term suggests *general directions* for its use,[12] and this on various levels. Thus the sense of the term "woman" generally restricts its use to that in referring to adult female humans; it *allows* its (mistaken, but still possibly successful) use in referring to men; and it *excludes* its use in referring to dreams (other than metaphorically). On the other hand, however, one is not so restricted when it comes to the use of names; no such general directions are provided for names, the notions they stand for being arbitrary and provided in the act of naming. Thus where terms with senses usually obtain those senses through use (though the senses of e.g. certain technical terms may be stipulated), names obtain their *use* by stipulation, a use which may become conventionalised. I can name *anything* "Andrew;" or, for that matter, I can name *nothing* "Andrew," and then proceed to use the term in keeping with my naming; this I cannot do with everyday terms that have sense.

Thus, among other things, when I say that the meaning of "bird-antelope" is the concept of a particular sort of gestalt-switch figure, or even if I were to *define* it as such, I am *not* saying that the meaning *is* the figure; and I *am* implying that the term can be used in referring to a variety of instances of the bird-antelope.[13]

Two terms are *synonymous* if they have the same sense (but not necessarily the same connotation), in which case it should most often be possible to exchange one for the other in a sentence in such a way that the resulting sentence has the same meaning as the original. Terms with sense may also be intersubstitutable in certain contexts if they have different senses but the same *referent(s)*, as may names. And a name may even be substituted for a referential term which is

---

[11] In this regard, cf. Strawson (1950), p. 110.
[12] Cf. ibid., p. 112.
[13] Cf. ibid., p. 117.

not a name (including what Russell calls 'denoting phrases') and vice versa if they both have the same referent. But in none of these cases are we speaking of a hard and fast rule; its application will depend on the nature of the original sentence and the context in which it is uttered. Similarly, whether or not the meaning of two terms is the same, or even whether a term has the same meaning when used on separate occasions,[14] depends in part on the context.

Naming vs. Referring vs. Describing

A name is a linguistic entity, more particularly a *term*. In some cases we may name objects ourselves, and in others an object's name may be determined by someone else or by convention. In either case naming, which involves reference, is to be distinguished from reference itself, which need not (and usually does not) involve naming. Further, we can say that naming consists in the original association[15] of a term with the concept (notion) of its referent *as* its referent. Naming paves the way for *using* the name by explicitly making the intentional connection between the name and its referent.

Also, as pointed out in the Gestalt Model and intimated earlier in this section, it is important to distinguish referring from describing, noting that an object may be successfully referred to despite being misconceived by both the person referring and his listener.

Identity and the Analytic vs. the Synthetic

On the present theory, as is in keeping with the everyday and general philosophical conception of identity, *two* entities are never identical. If *A* and *B* are identical, then *A* and *B* are *one and the same*. The figure of which the bird is an aspect is *identical* with the figure of which the antelope is an aspect, but the bird aspect is not identical with the antelope aspect nor with the figure. Thus identity is not a relation, it being necessary that a relation be between at least two *relata*; rather, the notion of identity marks the *absence* of relation, or, positively, marks that what is being considered is only *one* entity.

---

[14] Consider e.g. the use of the term "nice," treated in *this volume*, pp. 76–77.
[15] For a treatment of this concept of association in another context, see Dilworth (2007), pp. 246–247.

And, as in the Gestalt Model, whether saying that A is the same as or identical with B is analytic (tautological) or synthetic will partly depend on the situation in which it is being said, including the state of knowledge the speaker assumes in his listener. Thus in certain cases, such as that of naming, such an expression may be analytic, while in others, where the listener is being informed that an A and a B of which he has independent concepts are in fact the same entity, the assertion may be synthetic.

The Importance of the Context

Quite generally, we see from consideration of the Gestalt Model how important the context is with regard to acts of reference. Much depends on the states of knowledge of both the speaker and the listener, especially as regards that of the other. If the speaker believes he can safely assume particular knowledge on the part of the listener, or that he has a particular conception of reality, then he may feel free to express himself in a way which would be misleading were the listener not to have that knowledge or that conception.[16] Though the present effort is not an investigation into the psychological aspects of linguistic reference, we must nevertheless not lose sight of the fact that language is intimately related to thought, and that thoughts are had by persons.

Equivalence

One can speak of many sorts of things as being equivalent, and the notion is often used in the context of attaining some end or performing some function, suggesting that the end can still be reached or the function performed if one entity is *substituted* for another. Thus certain actions may be equivalent – running twelve minutes is equivalent

---

[16] Cf. Strawson (1950), p. 120: "What in general is required for making a unique reference is, obviously, some device, or devices, for showing both *that* a unique reference is intended and *what* unique reference it is; some device requiring and enabling the hearer or reader to identify what is being talked about. In securing this result, the context of utterance is of an importance which it is almost impossible to exaggerate; and by 'context' I mean, at least, the time, the place, the situation, the identity of the speaker, the subjects which form the immediate focus of interest, and the personal histories of both the speaker and those he is addressing."

to walking one hour, when it comes to the burning of calories. Or more abstract entities may be equivalent: the Swedish *Fil. dr* degree is equivalent to the British Ph.D. *Synonymity* is here seen as a form of equivalence, as is the above-mentioned intersubstitutability of referential terms with the same referents. And just as certain things more generally may be intersubstitutable as regards the performance of some functions but not others, terms may be synonymous or otherwise equivalent in some contexts but not in others. In each case however there underlies the idea of intersubstitutability, either in principle or in practice, and this is the sense in which "equivalence" is to be used on our theory.

Equality

On the present theory, equality, like equivalence, is a relation, while identity is not. Equality, more particularly, is a relation which holds between concepts (senses) or properties which have the *same numerical value*, which means they can be represented by the same *number*. Thus when we say (taking an example from Frege) $1 + 1 = 6/3$, what are equal are the *concepts* on each side of the equation, in that they have the same numerical value and can both be represented by the number 2. $1 + 1$ and $6/3$ are not identical on the present theory, for e.g. the former involves addition and the latter does not.

The notion of equality is a bit peculiar, however, in that we can also say that $2 = 2$, where there is only one concept (here taking numbers to be concepts) which is being repeated: the 2 being referred to on the one side of the equals sign is *the same entity* as the 2 being referred to on the other. On the other hand, however, we should not say that 2 is equal to itself, but rather is identical with itself. The referent(s) of whatever is separated by an equals sign may *potentially* be two aspects of the same entity which, as it turns out in the present case, are but one aspect.

Consideration of the nature of equality can be extended beyond mathematics to physics, where in the case of physical laws expressed as equations what are equated are properties (quantities) rather than concepts. Thus, for example, Ohm's law states that in a closed circuit voltage is equal to current times resistance: $V = IR$. Voltage is not being said to be *the same as* current times resistance. What is being said to be the same in this case are the (variable) numbers in terms of which voltage and current times resistance can be represented. Thus, in general, to say that A and B are equal is not to say that A and B are identical, but rather that they have the same, possibly variable, numerical value (which may, in certain contexts, also make them *equivalent*);[17] and in mathematics equality is a relation between concepts, while in physics it is a relation between properties.

Let us now see if our theory will help us understand a classic form the issue of identity and reference has taken in the philosophy of language, namely that revolving around Frege's treatment of the subject.

4. FREGE'S CONCEPTION OF IDENTITY

The notion of identity ostensibly plays a central role in Frege's philosophy of language. On the present theory however (and in fact independently of it) what Frege considers to be identity is not identity, or ought not be conceived of as identity, but in one case equivalence, and in another equality.

Equivalence

At the beginning of 'Über Sinn und Bedeutung' (1892), Frege states that in his *Begriffsschrift* of 1879 he assumed identity[18] to be a relation between names or signs of objects. If we refer to the *Begriffsschrift*, we see that he there provides the following characterisation:

$$\vdash (A \equiv B)$$

is to mean: *the symbol A and the symbol B have the same conceptual content, so that A can always be replaced by B and conversely.*[19]

Noting first that the term "content" ("*Inhalt*") here (counterintuitively) plays the part of what will later become the term "referent"

---

[17] In this regard cf. Wittgenstein (1921), § 6.24.
[18] Frege's term is "*Gleichheit*," i.e. "*equality*," but he expressly uses it "in the sense of identity": Frege (1892), p. 75n. His reason for doing so is given in the following subsection.
[19] Frege (1879), p. 12; emphasis added.

(*"Bedeutung"*) in 'Über Sinn und Bedeutung,'[20] we see that this conception bears a close resemblance to Leibniz' various formulations of his '*Eadem sunt ...*' doctrine, which Frege explicitly embraces as his own definition of identity in his *Grundlagen der Arithmetik* of 1884.[21] The formulation there is: *Eadem sunt, quorum unum potest substituti alteri salva veritate*, which may be rendered into English as: "Those are the same which can be substituted one for the other with preservation of truth."

On first reading, this appears a peculiar doctrine. Normally, as on our theory, one considers entities to be the same as themselves and different from other entities. But this cannot be what is meant by "the same" here, for Leibniz is speaking of the substitutability of "one for the other."[22] Quite generally it makes little sense to speak of substituting something for itself, for to claim that a substitution has taken place implies that there has been an *exchange* of entities – were the first entity to remain, or be removed and then replaced, we should say that no substitution has occurred. It would thus seem that another notion may be intended, one associated with identity, yet distinct from it.

Applying our theory here, noting first that both Leibniz and Frege use the notion of *substitution*, we should of course say that what they are talking about is not identity, but *equivalence*, or more particularly when it is a matter of terms, *linguistic* equivalence, which may hold in certain contexts if the terms have the same referents or the same senses – the latter making them *synonymous*. In these terms then, Leibniz' doctrine is saying that referential terms that can be substituted for one another without affecting the truth of the statement in which they occur have the same referent(s).[23] In this way our theory solves the problem of how to conceive of substitutability in a particular context that arises in the work of both Leibniz and Frege in which

---

[20] In this regard see e.g. Angelelli (1976), p. 156, or Caton (1976), pp. 169–170.

[21] § 65. Cf. also Frege (1892), p. 80.

[22] The doctrine seems even more peculiar on a second reading, for it says the converse of what Leibniz apparently intends it to say.

[23] Cf. Church (1956), p. 300, where it is similarly suggested that what Leibniz should have said is that *things* are identical if the *name* of one can be substituted for that of the other.

their own characterisations are counterintuitive, namely by *not* seeing it as an instance of identity, but of equivalence.

A quotation from Leibniz may add further support to this claim. Leibniz says:

That *A* is the same as *B* signifies that the one can be substituted for the other, *salva veritate*, in any proposition whatever. ... Thus, Alexander the Great and the king of Macedon who conquered Darius. Likewise, triangle and trilateral can be substituted for one another.[24]

Noting first that Leibniz is here expressing the *converse* of the *Eadem sunt* ... doctrine as cited in Frege's *Grundlagen*, we remark that on the present theory one cannot substitute Alexander the Great for the king of Macedon, for they are one and the same entity, and one cannot substitute an entity for itself. What may be substitutable one for the other however are the terms "Alexander the Great" and "the king of Macedon," or perhaps the notion 'Alexander the Great' and the concept 'the king of Macedon.' But in neither of these cases are the entities in question being claimed to be identical, but to be equivalent in a certain respect. And the same may be said of the terms "triangle" and "trilateral."

It may also be pointed out that Russell, too, employs this mistaken way of thinking in his 'On Denoting,' where says: "If *a* is identical with *b*, whatever is true of the one is true of the other, and either may be substituted for the other in any proposition without altering the truth of falsehood of that proposition."[25] Here, in repeating Leibniz and Frege, Russell is both conflating terms with their referents and taking distinct entities to be identical.

The above is not to say that the notion of identity is not relevant in the present case. On the standard construal the entities in question are taken to be terms that have the *same* referent. Furthermore, on our theory we might say quite generally that equivalent entities are to be able to play the *same* role or perform the *same* function in some context or other. So in this case the notion of identity appears in its own right, but not in the way either Leibniz or Frege takes it to.

As regards equivalence, what is perhaps of greatest interest here is the notion of *linguistic* equivalence, due to its relevance to Leibniz

---

[24] Quoted in Curley (1971), p. 499.
[25] Russell (1905), p. 98.

and Frege, and subsequently to modern philosophy. On our theory referential terms may be divided into those which have meaning, and which thus may potentially be synonymous, and those which do not (names), and terms of both sorts may be intersubstitutable in some contexts but not in others. What Frege is advocating (quite mistakenly, as he realises in 'Über Sinn und Bedeutung') in the *Begriffsschrift* quotation, on the other hand, is that *all* referential terms that have the same *referent* should be interchangeable in *any* context. And a fair interpretation of both Leibniz and Frege would be that, in speaking of the intersubstitutablility of terms that have the same referents, they are adopting a *referential theory of meaning*. That is, they are taking the *referents* – at least of referential terms, including names – to be their meanings.

Equality

In the first volume (1893) of *Grundgesetze der Arithmetik*, Frege compares his characterisation of identity there with that of the *Begriffsschrift*:

Instead of the three parallel lines I have chosen the ordinary sign of equality because I convinced myself that it is used in arithmetic to mean the very thing that I wish to symbolise. In fact, I use the expression "equal" to mean the same as "coinciding or identical with," and this is just how the sign of equality is actually used in arithmetic.[26]

Here we pause to note that on our theory, as is in keeping with everyday thinking, "coinciding with" does not mean the same as "identical with." *A* can coincide with *B* without being identical with *B*; and it is debatable whether we normally conceive of entities as coinciding with themselves.

Frege gives a more detailed account of his conception of equations as expressing identities in 'Funktion und Begriff' from 1891. There, he argues against those who would suggest that e.g. 1 + 1 and 6/3 (our earlier example) are equal but not the same, saying that they are confusing the signs with the entity signified. (Note that he doesn't take up the question of whether the *senses* are being confused with the entity signified.) Thus, he says, the equals sign of arithmetic

---

[26] Frege (1893), pp. 120–121.

expresses strict identity, since the right- and left-hand sides of an equation signify the same thing, i.e. have the same *referent* – in our example, the number 2. Should one respond by suggesting that $1 + 1$ is a sum, while $6/3$ is a quotient, Frege would reply that $6/3$ is the number which, when multiplied by 3, gives the result 6,[27] i.e. it is the number 2, which is the same number as results from adding 1 to 1.

However, apart from the fact that no reference (or at most an indirect reference) is being made here to what the operations characterised in the equation will result in, viz. the number 2, more is involved in any case than the statement that two signs have the same referent. On our theory, we should first distinguish between the terms "$1 + 1$" and "$6/3$" and the concepts '$1 + 1$' and '$6/3$,' and note that neither the terms *nor the concepts* are identical. What is identical is the number *that results from performing the operations* indicated on each side of the equation. Here we have a situation similar to that where in 'Über Sinn und Bedeutung' Frege introduces the notion of sense in order to account for the informativeness of identity statements *per se*. There he says:

If the sign "a" differs from the sign "b" only as an object (here by its shape) but not by its role as a sign, that is to say, not in the manner in which it designates anything, then the cognitive significance of "a = a" would be essentially the same as that of "a = b," if "a = b" is true.[28]

And his conclusion is that for "a = b" to be informative the signs "a" and "b" must therefore differ in sense,[29] or, as we should say so as to include names, they should stand for different concepts.

"$1 + 1 = 2$" is informative; it differs in cognitive significance from "$6/3 = 2$." Granting that the *result* of adding 1 to 1 is identical with that of dividing 6 by 3, i.e. is the number 2, "$1 + 1$" is nevertheless not just another way of writing "2," and "$1 + 1 = 2$" tells us more than simply that the symbol on the left of the equals sign has the same referent as that on the right.

---

[27] Frege (1891), p. 23.

[28] Frege (1892), p. 76; in keeping with his conviction that equality and identity are the same, Frege is here using the equals sign as a sign for identity.

[29] Angelelli (1976, p. 156) suggests that this well-known argument of Frege, a similar one of which also appears in Ch. I, § 8 of the *Begriffsschrift*, was not new, Frege's distinction between *Sinn* and *Bedeutung* being essentially that of earlier authors between *rationes* and *res*.

In his treatment of this example, Frege is reverting to the referential theory of meaning he employed in his *Begriffsschrift*[30] in a situation in which, on his more sophisticated way of thinking in 'Über Sinn und Bedeutung,' requires the introduction of the notion of *sense*. What the equation tells us is that the sense or concept on its one side bears a certain relation to that on the other: *not* that they have the same referent, but that they are *equal*.

Thus, on our theory, it is the concept of adding 1 to 1 which is equal to the concept of 2, due to both concepts' having the same numerical value, namely, 2. In the limiting case there may be but one concept expressed twice, as e.g. in the equation "2 = 2." But to say that 2 = 2 is not to say that 2 is identical with 2, or that 2 is identical with itself. It is rather to say that the concept '2' has the same numerical value as the concept '2,' viz. 2. (Here we note that, on our theory, '2' and 2 are identical, i.e. the concept '2' and its numerical value are the same entity.)

Thus Frege is mistaken in thinking that equality is (or, we should say, ought to be conceived of as) the same thing as identity. And the problems that arise from his mistake, such as the idea that equations should then be tautologies, are solved by the present theory's depicting equality not to be the same as identity, but to be a *relation* which holds between concepts (senses) or properties which have the same numerical value.

5. QUINE, KRIPKE, AND IDENTITY'S BEING A RELATION

Having considered seminal texts in the philosophy of language which on our theory mistakenly conceive of equivalence and equality as identity, we might go on to look at what some more recent commentators have to say about the notion of identity. More particularly, we shall investigate the widely held view, largely derived from Frege, that identity is a relation.

At the beginning of 'Über Sinn und Bedeutung,' where Frege concerns himself with the nature of identity, he asks. "Is identity a relation? A relation between objects? Or between names or signs of

---

[30] Cf. Caton (1976), p. 177.

objects?"[31] He never answers the first question however, and throughout his work he simply assumes identity to be a relation. Virtually all logico-linguistic philosophers have followed Frege in this respect; and of course the terminology of identity's being a relation has become well established in formal logic.

The counterintuitiveness of this view suggests that those who hold it should provide some argument or rationale for it; and this in fact some of them do. Quine, for example, taking identity to be a relation between a named object and itself, says: "What makes identity a relation ... is that '=' goes between distinct occurrences of singular terms ..., and not that it relates distinct objects."[32] Quine, following Frege, is here using the equals sign as short for "is the same object as," and so his argument is from natural language and not, as might first be thought, from formal languages.

Granting with Quine that identity is not a relation due to its relating distinct objects, since distinct objects ought not be conceived of as being identical, we might look at the positive part of his reasoning more closely. While his argument is from linguistic convention, it may be noted that he is not saying e.g. that identity is a relation because, in language, it is normally called or considered a relation. Rather, he is saying that identity is a relation because its linguistic representation has the form of a relation. But if one is allowed to argue from linguistic form one might just as well say that identity is a relation between *two* things, since its linguistic representation involves two (token) terms. Arguing in this manner can lead to such patent non sequiturs as that ideas are physical, since their linguistic representations are, or that Paris cannot be very large, since its linguistic representation is not. Quine is committing a variation on the very fallacy he warns against in this context (and which has already been expressed by Frege),[33] namely that of confusing sign and object. Language is a medium we use to convey information concerning what is the case, not to determine it. Furthermore, though less important, in being an argument from convention Quine's reasoning is open to the vicissitudes of any argument from convention, and is equally as circumstantial.

---

[31] Frege (1892), p. 75.
[32] Quine (1960), p. 116.
[33] Cf. Frege (1891), p. 23, treated on p. 249, above.

Kripke, too, believes identity to be a relation between an entity and itself, and also provides an argument which is connected to everyday language. He says:

[S]ome philosophers have thought that a relation, being essentially two-termed [*sic*], cannot hold between a thing and itself. This position is plainly absurd. Someone can be his own worst enemy, his own severest critic and the like. Some relations are reflexive, such as the relation 'no richer than.' Identity ... is nothing but the smallest reflexive relation.[34]

To say of someone that he is his own worst enemy is a figurative way of saying e.g. that his advancement is hindered by his own doings; and it is difficult to see how the possibility of someone's being in such a position should have any bearing on the question whether identity is a relation. If we are to take figurative expressions at face value, as we must in order for Kripke's argument to get off the ground, then we could just as well argue that our being able to say of someone that he is not himself today shows that a person need not be identical with himself. Even if we were to take the particular locutions employed by Kripke as they stand, why should their occasionally being true indicate that there can exist a relation between a person and himself, and not rather that a person can have a particular sort of attribute (e.g. that of 'being his own worst enemy')? And while someone's being no richer than someone else might perhaps be thought to constitute a relation between them, Kripke has not shown that there is also a relation in the case where a person is said to be no richer than himself, and so his reference to this notion does not really support his conclusion that identity itself is a relation of the same sort.

That neither Quine nor Kripke has provided a convincing argument for identity's being a relation (or, as we should say, for identity's best being conceived of as a relation) does not mean that it is impossible to provide such an argument.[35] But if we survey the various contexts in which the idea that identity is a relation manifests itself, we find that what leads one to think of a relation is the presence

---

[34] Kripke (1980), p. 108n.

[35] Another such argument is given by Benson Mates (1972, pp. 156–157). Suffice it to say that Mates takes his opponent to hold that identity is not a relation because there is nothing literally *between* a thing and itself, rather than that it is not a relation because relations always involve at least two *relata*.

of entities which, while highly relevant to the situation, are not themselves identical. They may be names or other terms used in referring to one and the same thing; or they may be the concepts or senses the terms stand for; or they may be different aspects of the entity in question. It is here suggested that the thought that identity is a relation is prompted by confusing relations between the above sorts of entities – relations which may include those of equality and equivalence – with identity itself. Once one sets these relations aside, however, and considers identity *per se*, it is difficult to see that there is any relation left.

6. FURTHER APPLICATIONS

The viability of the present theory can be further supported by applying it to various well-known conundrums in the philosophy of language concerning the notions of identity and reference.

The Morning Star and the Evening Star

The morning star/evening star problem is but one expression of the problem which led Frege in 'Über Sinn und Bedeutung' to introduce the notion of sense.[36] Given a Leibnizian conception of the synonymity of referential terms there should be no epistemologically relevant difference between saying that the morning star is the morning star and the morning star is the evening star: both should be tautological since "morning star" and "evening star" are to be interchangeable due to their having the same referent.

On the present theory, on the other hand, terms may be interchangeable in some contexts but not in others, and in some cases because they have the same referent(s) and in others because they have the same sense. Thus in cases where two terms have the same referents but differ in sense they may or may not be interchangeable, depending on the context, including the wording of the relevant sentence.

As regards the analytic/synthetic question, if one person is speaking of Venus as the morning star, and another person only knows of Venus as the evening star, then his being told that the morning star is

---

[36] Cf. Frege (1892), p. 76.

the evening star (i.e. that the morning star and the evening star are the same thing)[37] could be analytic. On the other hand, if an ancient astronomer were to learn that the bright 'star' he saw in the morning was the same heavenly body as the one he saw in the evening, then "the morning star is the evening star" would be synthetic. Thus we should say that "the morning star is the evening star" may be analytic or synthetic, depending on the context in which it is uttered. "The morning star is the morning star" (i.e. "the morning star is identical with itself"), on the other hand, should be tautological in virtually any context. In none of the above examples, however, are the terms interchangeable.

Note (independently of the analytic/synthetic question) the possibility of giving two interpretations to "the morning star is the evening star." On the one interpretation it would be to say that the morning star and the evening star are the same thing (our parenthetical insertion above); on the other it would be to say that the morning star and the evening star are *different aspects of* the same thing. In terms of the Gestalt Model the first interpretation would be like saying that the Gestalt Model for the present theory is the same thing as Hanson's bird-antelope; and the second would be like saying that the bird and antelope are different aspects of the same figure. Thus the sentence "the morning star is the evening star," apart from being an unnatural form of expression, does not in itself provide enough information for the person hearing it to determine what the speaker intends in this regard.

We further note that this construction is also ambiguous in another respect, namely that it doesn't make clear for the listener whether the speaker by e.g. "morning star" simply means Venus or more particularly Venus *in the morning*. Venus in the morning is not the same thing as Venus in the evening, and if in saying that the morning star is the evening star one meant that they *were* the same, then the statement would be false, just as it would be false that some particular bird and some particular antelope are the same thing.

---

[37] Cf. Linsky (1951), p. 89: "Only the logician's interest in formulas of the kind 'x = y' could lead him to construct such sentences as 'The Morning Star = The Evening Star' or 'Hesperus = Phosphorus.' Astronomers and mythologists don't put it that way." The reason this awkward form of expression is used has to do with the syntax of the sort of language chosen to express the relevant concepts, as will be considered below.

The Present King of France

This problem has been expressed by Russell, as follows:

By the law of the excluded middle, either "A is B" or "A is not B" [rather, "it is not the case that A is B"] must be true. Hence either "the present King of France is bald" or "the present King of France is not bald" ["it is not the case that the present King of France is bald"] must be true. Yet if we enumerated the things that are bald, and then the things that are not bald, we should not find the present King of France on either list.[38]

Russell is here assuming that, in ordinary language, what is to constitute a sentence is determined wholly by grammatical form, that all sentences should be meaningful, which should make them either true or false (his reference to the law of the excluded middle), that their meanings are independent of the context in which they are uttered, and that in order to be meaningful their 'referring terms' must have referents (or, as it turns out, be explicitly *denied* referents).

That the example of the present King of France is a problem for Russell given his presuppositions is clear, for if there is no such entity how can sentences in which the term occurs be true or false given his view that referential terms must have referents? Russell's efforts here may thus be seen as having the aim of making it possible meaningfully to use ordinary-language terms the referents of which do not exist, despite the fact that they should be meaningless on his theory of meaning, i.e. despite the fact that on his theory it is impossible to refer to that which does not exist other than explicitly to claim it to exist or not to exist.

Russell's 'solution' is to distinguish between what he terms primary and secondary occurrences of "denoting phrases" and to reconstruct one of the sentences quoted above in such a way that whether or not "the present King of France" has a primary or secondary occurrence should be clear from the syntax of the reformulated sentence. Thus, on Russell's interpretation, the first sentence in the above quote would always be false, but the second would be false if the occurrence of "the present King of France" is primary, and true if it is secondary. Thus the second sentence – "it is not the case that the present King of France is bald" – is to be susceptible of reformulation

---

[38] Russell (1905), p. 98.

as two essentially different sentences: either "there is an entity which
is the present King of France and is not bald," in which "the present
King of France" has a primary occurrence and which is false; and "it
is false that there is an entity which is the present King of France and
is bald," in which "the present King of France" has a secondary
occurrence and which is true.[39]

But Russell's distinction between primary and secondary occurrences of denoting phrases does not solve his problem, but only puts
it into a more specific form. Thus his problem now becomes that of
determining whether "the present King of France" is occurring in a
primary or secondary way in "it is not the case that the present King
of France is bald;" and Russell has not indicated how to do this. He
has depicted an ontological distinction, namely that between primary
and secondary occurrences of denoting phrases, but no epistemological distinction allowing us to recognise the ontological distinction. In
other words, he has provided no way of determining whether ordinary sentences which deny properties of non-existent entities are true
or false.

On our theory, on the other hand, we first distinguish between the
meaning and the (possible) referent of "the present King of France"
and note that while the phrase has a clear meaning, as used at present
in an everyday situation it would have no referent. Thus the *statement*
that might be expected to be made using the sentence "the present
King of France is bald" *would* involve the making of a reference, but
an *unsuccessful* reference, due to the fact that there is no object being
referred to.

On the other hand, however, the sentence *could* be used in making
a statement that involves a successful reference, if, for example, it is
being uttered at a time when France had a monarchy, or in the context of a play. In the latter case the referent would exist *as intended*,
i.e. as a character in a play, while it would not exist as a thing. And
the question as to whether the statement the sentence was being used
to make is to be considered true, false or nonsense would have to be
considered in each instance with respect to the context in which it is
being made.

---

[39] Ibid., pp. 102–103.

Linsky's Case

According to Leonard Linsky:

> [S]aid of a spinster that 'Her husband is kind to her' is neither true nor false. But a speaker might very well be referring to someone in using these words, for he may think that someone is the husband of the lady .... Still, the statement is neither true nor false, for it presupposes that the lady has a husband, which she has not. This last refutes [the] thesis that if the presupposition of existence is not satisfied, the speaker has failed to refer. For here that presupposition is false, but still the speaker has referred to someone, namely, the man mistakenly taken to be [the woman's] husband.[40]

Linsky draws the same conclusion regarding the present case as we would on our theory, namely that a (successful) reference has been made. But the reason he gives for his claim that the statement in question should be neither true nor false (that it presupposes a falsehood) does not support that conclusion, but rather that the statement should be *false*. On our theory, on the other hand, one of at least two different statements may be intended, which may be either true *or* false. Thus, intended as involving a reference to a man who turns out not to be the lady's husband it could be true – or false – but perhaps in need of qualification (e.g. by the speaker's saying, "by 'her husband' I meant this man"). In the present context what is important is that the listener be able to understand what object is being referred to, not that the characteristics ascribed to the object in order to delimit it actually belong to it.

Also, it may be pointed out that Linsky fails to distinguish between successful and unsuccessful reference, as we do on our theory. While on the one hand a person who is the lady's husband does not exist, on the other the person being referred to does, and Linsky fudges this distinction when he says that the presupposition of existence is not satisfied. The person to whom the concept correctly applies does not exist, but the object of the act of reference does, and as long as the listener is able to delimit that object, the reference is successful. Should there be no one being referred to, however, given that some real person is intended by the speaker, the reference would be unsuccessful.

---

[40] Linsky (1951), p. 80.

Fictional and Mythological Characters; the Round Square

As suggested earlier, while on our theory the (epistemological) *object* of any successful reference must exist at least as an intentional object, there need not always be a(n ontological) *thing* of which the object is a manifestation. Thus reference can successfully be made to fictional and mythological characters, as implied above. Hamlet's being an object of such a reference is dependent on his being conceived in a realm in which he does exist: quite broadly, that of fiction; and, similarly, successful reference to Zeus is dependent on his being conceived as existing in the realm of myth.

On the other hand, one cannot, even unsuccessfully, refer (other than problematically) to such an entity as a round square, since it is a priori impossible to form a concept of such an object. (It would perhaps be like attempting to go fishing using tackle that is impossible to assemble.) In other words, the term "round square" is nonsense, and the attempt to use a nonsensical term in referring itself results in nonsense.

7. TO CONSTRUCT A LOGICALLY PERFECT LANGUAGE

Should we seek the origin of the various modern problems of identity and reference, we would find that they all stem from the belief on the part of the founders of the logico-linguistic tradition in the idea of constructing an ideal language, and the characteristics they believed such a language should have. For Frege, following Leibniz, as for Russell, Wittgenstein, Carnap and many others, the philosophical passion at the end of the nineteenth and beginning of the twentieth centuries was one for the idea of, and in some cases the creation of, just such a language. This language was to be free from metaphysics, and to involve no vagueness or ambiguity, such that truths which were *certain*, once discovered, could be preserved and communicated to others.[41] Such a language was to be, more particularly, a *logically perfect language*,[42] its syntax being that of formal logic, and

---

[41] This approach, starting with Leibniz, may be seen as part of the reaction to the displacement of non-materialistic philosophy – particularly the logico-linguistic aspect of Aristotle's philosophy – with the materialistic philosophy of modern science. In this regard see Dilworth (2007), pp. 200–203&n. and 263.

[42] Cf. Frege (1892), p. 86.

its semantics consisting in the first instance in the referents of the language's terms, and later in the constant functional relations between the terms and their referents, there being however no essential difference in the two approaches, the terms "reference" and "referent" being used interchangeably. The semantics of the language was thus that of Leibniz and Frege, synonymity (interchangeability *salva veritate*) consisting in terms' having the same referents, a state of affairs which was to include propositions or statements, which were also to have referents, and which were to be conceived of as *sentences*.[43] Among other things, this means that definitions in the language, rather than giving the meanings or senses of terms, are to indicate their *extensions*, and a term may thus be considered properly defined when and only when its extension is strictly delimited. Another way of putting this is to say that a logically perfect language is a purely referential language, in which the meanings of its terms are their referents or the relation between the terms and their referents; and in this way all terms, including sentences, become names. Here certainty is to be assured by the terms' referents being restricted so as to include only indisputable *sense-data*, which are empirical, while the syntax of the language, being formal logic, should allow the mechanical or *algorithmic* determination of non-empirical truths.

This is one way of describing the basis of logico-analytic philosophy. Another might be to say that it is the doctrine that philosophical matters are best handled by expressing them in terms of a formal calculus. It is along this line that Ryle, for example, says:

This doctrine is the doctrine that philosophical disputes can and should be settled by formalising the warring theses. A theory is formalised when it is translated out of the natural language ... in which it was originally excogitated, into a deliberately constructed notation, the notation, perhaps, of *Principia Mathematica*. The logic of a theoretical position can, it is claimed, be regularised by stretching its non-formal concepts between the topic-neutral logical constants whose conduct in inferences is regulated by set drills. Formalisation will replace logical perplexities by logical problems amenable to known and teachable procedures of calculation.[44]

---

[43] In this regard, cf. Strawson (1950), Sect. II.

[44] Ryle (1953), p. 125.

But, as suggested, this is just another way of saying that philosophical matters should be expressed in a logically perfect language; or rather that, since there is no such language, they should be expressed in terms of one's favourite candidate for the syntax of such a language. Another way of putting this is that *reason* should be constrained, to the extent possible, by some logical calculus.[45]

Despite the fact that a logically perfect language was never forthcoming, and the use of formal logic in philosophy does not solve problems but only gives rise to them, the idea itself nevertheless lived on through a focus on its syntax, i.e. on formal logic. This was undoubtedly at least partly due to its attraction to puzzle-solving logicians and its partial applicability to mathematics.[46] And, apparently without any thought being given to the matter, its attributes were assumed, ideally, to exist in real language (and actually to *constitute* the conceptual structure of modern science!),[47] such that, among other things, the concepts expressed in the context of a real language could actually be *better* expressed using the syntax and referential semantics provided by a formal calculus, as suggested in the Ryle quotation. Given the plethora of formal calculi that have been created, however, many of which, in application, are not compatible with one another, the question arises as to *which* calculus is to be used. But this question is seldom broached, logico-linguists most often simply assuming whichever calculus they prefer, with no reason being given for their choice, nor even an explicit indication of which choice they have made.

Thus, assuming without argument a referential theory of meaning, which itself is counterintuitive, we see such efforts as those of Frege to establish that *sentences* in ordinary language also have referents, viz. the True and the False. And we see similar efforts on the part of Russell (treated above) to alter the syntax of sentences from ordinary language – sentences which *prima facie* appear nonsensical, but whose syntax *should* be acceptable in a logically perfect language (they should be 'well-formed formulae') – in such a way that their 'semantics' becomes acceptable, so that they may be construed as

---

[45] In this regard cf. Dilworth (2007), p. 67&n.

[46] In this general regard, cf. Strawson (1950), p. 121.

[47] Its doing so being presupposed by both logical empiricists and Popperians, as treated in the main text.

being true or false.[48] We also see attempts – already by Frege[49] – to solve the problem that this way of thinking gives rise to when applied to what have been termed *opaque contexts* in ordinary language.

Here we pause to note a way of thinking that is endemic to the analytic tradition (including ordinary-language philosophy), and which has been taken up with respect to the present volume in Appendix II, that is, that no distinction is made between the provision of a philosophical theory on the one hand and conceptual analysis or factual generalisation on the other. It would be one thing to posit as a theory that ordinary language or thought has the attributes of a logically-perfect language (or, with reference to the main text, that modern-scientific thinking has the form of the syntax of an ideal language), and then investigate the extent to which this is actually the case. But what is done on the part of logico-linguists is that its being so in all essentials is merely assumed, i.e. the *theory* is assumed implicitly, and then changes are made ad hoc whenever in application the theory runs into difficulty (as, for example, when Frege introduces the notion of sense, or when Russell distinguishes between denoting phrases the occurrence of which is to be primary as versus secondary, or more generally when a distinction is made between terms which are to be taken as referring and terms which are not[50]). Matters which should concern the underlying viability of the theory are treated as factual matters to be reasoned about, with reason itself being constrained by the syntax and semantics of some non-existent ideal language, while this presupposition is itself nowhere admitted or, most often, even recognised. This is part of the anti-metaphysical, anti-system-building ethos of twentieth-century analytic philosophy. Rather than have no metaphysics, the metaphysics is assumed without its even being realised that it is being assumed. Thus the effort made to deal with conceptual issues aims no higher

---

[48] Cf. Strawson (1950), p. 108.

[49] Cf. Frege (1892), pp. 82ff.

[50] Or, with reference to the philosophy of science, when Popperians make claims about science which cannot be expressed in terms of the Deductive Model; or when Carnap weakens the logical empiricist criterion of meaningfulness in order to accommodate theoretical terms (*this volume*, p. 132). This remark also applies to the set-theoretic conception of science (see ibid., pp. 115–116), which also involves the application of a formal system, though not one as might constitute the structure of an ideal language.

than to produce correct generalisations – generalisations which are, to the extent possible, in keeping with their being conceived of in terms of a logically perfect language.

Thus, implicitly presupposing reason to be limited to the dictates of what a logically perfect language will allow, we see that virtually all commentators on such questions as those regarding identity and reference, including Leibniz, Frege, Quine and Kripke, take them to be ontological: *Is* identity a relation? *Do* names have sense? *Is* "the morning star is the evening star" a tautology? But, as is in keeping with the introductory remarks of this appendix and has been touched on in passing throughout, like virtually all philosophical issues, the sorts of question treated here are not ontological, but, in a broad sense, *epistemological*, or, perhaps better, *conceptual*. It is not a matter of what identity and reference *are*, but how we might best *conceive* of them; the issue does not concern reality *per se* conceived of in a certain way, but *how* it is to be conceived. Reference to facts is relevant in showing the applicability of a theory, or in criticising it, but the higher philosophical aim is not to point out those facts but to make sense of them. And to do this one must provide a comprehensive, unified *theory* in which they can be accommodated – moreover, preferably a theory which is advanced *explicitly*.

Furthermore, what is missed by virtually all contributors to the ideal language/ordinary language debate, including Russell and Strawson, is that the aim of the efforts of the ideal-language philosophers and the ordinary-language philosophers is not the same, and the treating of the issue as concerning factual matters only hides this fact and results in a talking at cross purposes. For the ordinary-language philosophers (other than Wittgenstein and his followers) such questions *are* only factual – wholly dependent on the nature of ordinary language – while for the ideal-language philosophers they involve the application of an implicit theory. Thus Strawson cannot fairly criticise Russell's 'theory of descriptions' as though it were simply intended to provide a correct general description of ordinary language.[51] What

---

[51] Cf. e.g. Strawson (1950), p. 106: "Russell's Theory of Descriptions ... is still widely accepted among logicians as giving a correct account of the use of ['denoting phrases' occurring as the grammatical subjects of sentences] in ordinary language." That Strawson says "among logicians" gives a hint that his analysis should go deeper, as do his considerations in ibid., p. 121, referred to above.

Strawson should criticise is the *theory* which leads to Russell's particular treatment of ordinary language (and by this I of course do not mean his 'theory of descriptions' but his referential theory of meaning and the ideal language in which it is to be embedded). And Donnellan cannot point out a particular general fact regarding ordinary language and claim that it counts for or against what either Russell or Strawson has said as though there were not a fundamental difference in what Russell and Strawson are doing.[52] And to this we should add that contributions in this regard on the part of Strawson, Donnellan and others of the ordinary-language persuasion, while they succeed in pointing out certain generalisations that may be made regarding ordinary language, from the point of view of philosophical *theory* are only *critical* (or, potentially, confirmatory). These philosophers succeed in indicating weaknesses with regard to the applicability of the logico-linguists' underlying theory (without expressly directing their criticisms to it), but they do not advance an alternative (as is done in the present appendix).[53] Similarly, in the philosophy of science the contributions of Kuhn and Feyerabend are only critical (of the formal, ideal-language theory of science), for neither do they advance an alternative (as is done in this book).

In any case, an ideal language, as conceived in the tradition, is a very Spartan affair, where the lack of a notion of sense is only one of its lacunae as compared to ordinary language. Another missing element that is crucial, as was pointed out with regard to reference by Strawson as referred to above,[54] is the personal: it is *people* who use language. And people can and do use the same words in different ways in different situations. Furthermore, ordinary languages evolve

---

[52] Cf. e.g. Donnellan (1966), p. 283, where he says, "one or the other of the two views, Russell's or Strawson's, may be correct about the nonreferential use of definite descriptions, but neither fits the referential use;" and p. 297: "neither Russell's nor Strawson's theory represents a correct account of the use of definite descriptions." It was hardly Russell's aim to provide such an account.

[53] For the mistaken view that both Russell and Strawson are each advancing theories in the texts being referred to here, see e.g. the second quote in the previous footnote. In Russell's case, despite his claiming to be putting forward a theory regarding denoting phrases, what he is presenting is an *hypothesis* couched in the context of his ideal-language theory. And what Strawson is doing is presenting hypotheses concerning the nature of ordinary language.

[54] P. 234, n. 5.

through use. Not only the senses[55] but also the possible referents of terms can change with time.

This is not the case however with a logically perfect language. As mentioned above, in order to be able to afford the communication of certainty for which it is designed, the meanings (referents) of its terms must remain the same at all times and in all contexts. One way of building this into the way one speaks about such 'logically perfect' terms is to say that they themselves refer, so that the same (type) term will always have the same referent! In one of the most widely read philosophical works of the last fifty years,[56] Kripke devotes the major portion of his energy to arguing that names in ordinary language have this property of logically perfect words (individual constants in a calculus), ordinary names thereby being what he calls rigid designators. Like Russell as treated above, throughout the text Kripke presupposes – quite possibly unwittingly – that ordinary language *as a matter of fact* has the majority of the characteristics of a logically perfect language, but (unfortunately) not others; and he takes it as his task to show that, given his presuppositions, it has as many such characteristics (particularly when it comes to reference) as can reasonably be argued for.

Kripke's book is paradigmatic of the kind of work that has been most influential in the logico-linguistic tradition, particularly as regards its presupposing ordinary language already to have the basic structure of a logically perfect language. In this regard, where words rather than people are conceived of as doing the referring in the ideal language, and no distinction is made between a term and its use, such is assumed to be the case when it comes to ordinary language; where a distinction between the reference and referent of a term is not made in the context of the ideal language, neither is it made in the case of ordinary language; where identity is represented as a relation in the ideal language, it is conceived as being a relation in reality; where in the ideal language a term's meaning or sense may be taken to be a functional relation between the term and its referent (i.e. to be its *reference*), so is it taken in ordinary language. Where essence is conceived of in terms of necessary and sufficient conditions in the ideal

---

[55] Cf. above, p. 242, n. 14 and accompanying text.
[56] Kripke (1980).

language, so is it conceived in natural language. And where no distinction is drawn in the ideal language between the analytic and the a priori, or between the synthetic and the a posteriori, neither are such distinctions drawn with respect to ordinary language or philosophical thought.

One point stemming from all this that has to be stressed with regard to the logico-linguistic tradition as a whole is its *static* approach. Not only are the meanings (referents) of linguistic entities (in the best of cases) to be fixed once and for all, so is their conceptual status. Thus if a statement expressed by saying or writing a particular sentence is analytic or false in some context, the sentence itself is to be analytic or false wherever it appears[57] (unless an ad hoc modification says otherwise). Nor is any epistemologically relevant distinction made between ordinary language and mathematics (as in Frege's treatment of equality as being identity), the latter, as noted above, being more amenable to the application of notions intended for an ideal language – among other things, the meanings of its expressions being context-independent. The employment of this way of thinking with regard to real languages or thought I term *context-blindness*.

Another point concerns the application of the law of the excluded middle. Any instance of a sentence is to be analytic or synthetic (a posteriori) and nothing in between, and to be either true or false and nothing in between. This approach I call *box-thinking*.[58] Box-thinking is a constraint imposed by the syntax of the logically perfect language, while context-blindness is a reflection of the 'semantics' of the language, which, as mentioned, is to take the form of a set of timeless mathematical functions or the values of such functions. Both of these aspects of an ideal language were to be aids to clarity, but taking them from the conception of an ideal language and applying them to reality is precarious – and in fact misleading when it is

---

[57] Cf. Strawson (1952, p. 235) in this regard: "[T]he preoccupation with the ideal type of sentence explains the persistent neglect of the distinction between sentence and statement. For, in the case of sentences of the ideal type, the distinction really *is* unimportant. Such a sentence, whenever it is used, is used to make one and the same statement; the contextual conditions of its use are irrelevant to the truth or falsity of that statement." Regarding the distinction between sentences and statements or propositions, see also *this volume*, p. 73.

[58] Concerning context-blindness and box-thinking, see Dilworth (1992), pp. 207–208.

not done explicitly – and, as a matter of fact, virtually always leads to problems.

With this understanding, then, we can see that the problems of identity and reference have a particular similarity to the empiricists' and Popperians' problems in dealing with scientific progress treated in the main text, and that both sorts of problem result from the taking of a formal-logical approach in philosophy, or, in other words, from the unwitting presupposition that reason, whether philosophical or scientific, has the structure suggested by the syntax and semantics of an ideal language. In fact we can say that, apart from the actual construction of formal calculi, virtually the whole of logico-linguistic philosophy consists in grappling with the various problems that arise from taking reality (modern science, ordinary language, thought) to be in keeping with one or the other such calculus; and the purported solutions often consist in the ad hoc weakening or transcending of the requirements set by its employment.

That the formal approach involves abstraction is not itself a problem – the present view, as well as that of the main text, also involves abstraction. But in the weighing-off between too little and too much abstraction, the formal approach falls heavily on the side of too much. And the ultimate solution of the problems this approach gives rise to requires abandoning it in favour of an alternative that is less abstract.

For the above reasons we can also see a similarity between the *solutions* offered in the present book to the sorts of problem that arise in the philosophy of language and the philosophy of science respectively. Apart from the solutions' to both sorts of problem being exemplifiable by the Gestalt Model, they in each instance consist in the presentation of a comprehensive theory which deals with the issues explicitly *as* a theory. Furthermore, not only do these solutions not afford an algorithm, but, perhaps most important, they place thinking and acting individuals in the centre of considerations.

REFERENCES

Achinstein, P.

(1964) 'On the Meaning of Scientific Terms,' *Journal of Philosophy* **61**, 1964.

(1965) 'The Problem of Theoretical Terms,' *American Philosophical Quarterly* **2**, 1965.

(1968) *Concepts of Science*, Baltimore: Johns Hopkins Press, 1968.

Adams, E. W.

(1959) 'The Foundations of Rigid Body Mechanics and the Derivation of Its Laws from those of Particle Mechanics,' in L. Henkin et al. eds., *The Axiomatic Method*, Amsterdam: North-Holland, 1959.

Agazzi, E.

(1976) 'The Concept of Empirical Data,' *Formal Methods in the Methodology of Empirical Sciences,* M. Przełecki et al. (eds.), D. Reidel, Dordrecht: D. Reidel, 1976.

(1977a) 'Subjectivity, Objectivity and Ontological Commitment in the Empirical Sciences,' in R. E. Butts and J. Hintikka eds., *Historical and Philosophical Dimensions of Logic, Methodology and Philosophy of Science*, Dordrecht: D. Reidel, 1977.

(1977b) 'The Role of Metaphysics in Contemporary Philosophy,' *Ratio* **XIX**, 1977.

(1978) 'L'Objectivité Scientifique: Est-Elle Possible Sans La Mesure?,' *Diogène* **104**, 1978.

Alexander, P.

(1963) *Sensationalism and Scientific Explanation*, London: Routledge & Kegan Paul, 1963.

Allen, H. S. and Maxwell, R. S.

(1939) *A Text-Book of Heat*, Part I, London: Macmillan and Co., 1944.

Alston, W. P.

(1971) 'The Place of the Explanation of Particular Facts in Science,' *Philosophy of Science* **38**, 1971.

Andersson, J. and Furberg, M.

(1966) *Språk och Påverkan*, Stockholm: Bokförlaget Aldus/Bonniers, 1974.

Angelelli, I.

(1976) 'Friends and Opponents of the Substitutivity of Identicals in the History of Logic,' in Schirn ed. (1976).

Aristotle

The Works of Aristotle, vol. 1, W. D. Ross ed., London: Oxford University Press, 1928.

Ayer, A. J.

(1936) *Language, Truth and Logic*, N.Y.: Dover Publications, 1952.

Barton, A. W.

(1933) *A Text Book on Heat*, London: Longmans, Green and Co., 1944.

Bhaskar, R.

(1983) Review of the first edition of *Scientific Progress*, *Isis* **74**, 1983: 258–259.

Blackmore, J.

(1984) Review of the first edition of *Scientific Progress*, *Annals of Science* **41**, 1984.

Bocheński, I. M.

(1961) *A History of Formal Logic*, Notre Dame, Indiana: University of Notre Dame Press, 1961.

Boltzmann, L.

(1896) *Lectures on Gas Theory*, S. G. Brush tr., Berkeley: University of California Press, 1964.

Bridgman, P. W.

(1936) *The Nature of Physical Theory*, Princeton: Princeton University Press, 1936.

Buchdahl, G.

(1964) 'Theory Construction: The Work of Norman Robert Campbell,' *Isis* **55**, 1964.

Burnet, J.

(1914) *Greek Philosophy – Part I: Thales to Plato*, London: Macmillan and Co., 1920.

Burtt, E. A.

(1924) *The Metaphysical Foundations of Modern Physical Science*, London: Routledge & Kegan Paul, 1980.

Campbell, N. R.

(1920) *Physics: The Elements*, Cambridge: Cambridge University Press, 1920; reissued as *Foundations of Science*, N.Y.: Dover Publications, 1957.

(1921) *What is Science?*, N.Y.: Dover Publications, 1953.

(1928) *An Account of the Principles of Measurement and Calculation*, London: Longmans, Green and Co., 1928.

(1938) 'Symposium: Measurement and its Importance for Philosophy: Part I,' *Proceedings of the Aristotelian Society, Supplementary Volume XVII*, 1938.

Carnap, R.

(1936) *Testability and Meaning*, New Haven: Whitlock's, 1954.

(1956) 'The Methodological Character of Theoretical Concepts,' in H. Feigl and M. Scriven eds., *Minnesota Studies in the Philosophy of Science* I, Minneapolis: University of Minnesota Press, 1956.

(1966a) *Philosophical Foundations of Physics*, N.Y.: Basic Books, 1966.

(1966b) 'Probability and Content Measure,' in Feyerabend & Maxwell eds. (1966).

Caton, C. E.

(1976) '"The Idea of Sameness Challenges Reflection",' in Schirn ed. (1976).

Caton, C. E. ed.

(1963) *Philosophy and Ordinary Language*, Urbana: University of Illinois Press, 1963.

Church, A.

(1956) *Introduction to Mathematical Logic*, vol. 1, Princeton: Princeton University Press, 1956.

(1951) 'The Need for Abstract Entities in Semantic Analysis,' in Copi & Gould eds. (1967).

Clark, M. E.

(1989) *Ariadne's Thread*, N.Y.: St. Martin's Press, 1989.

(1991) 'Rethinking Ecological and Economic Education: A Gestalt Shift,' in Costanza ed. (1991).

(1992) 'Critiquing the "Rational" Worldview: The Parable of the Tree-Trunk and the Crocodile,' in L. O. Hansson and B. Jurgen eds., *Human Responsibility and Global Change*, Gothenburg: University of Gothenburg, 1992.

Cohen, M. and Nagel, E.

(1934) *An Introduction to Logic and Scientific Method*, N.Y.: Harcourt, Brace & World, 1934.

Colodny, R. G. ed.

(1965) *Beyond the Edge of Certainty*, Englewood Cliffs, N.J.: Prentice Hall, 1965.

(1966) *Mind and Cosmos*, Pittsburgh: University of Pittsburgh Press, 1966.

Copi, I. M. and Gould, J. A. eds.

(1967) *Contemporary Readings in Logical Theory*, N.Y.: Macmillan, 1967.

Costanza, R. ed.

(1991) *Ecological Economics: The Science and Management of Sustainability*, N.Y.: Columbia University Press, 1991.

Cranston, M.

(1967) 'Francis Bacon,' in vol. 1 of Edwards ed. (1967).

Curley, E. M.

(1971) 'Did Leibniz State "Leibniz' Law"?,' *Philosophical Review* **80**, 1971.

Daly, H. E.

(1987) 'The Economic Growth Debate: What Some Economists Have Learned, But Many Have Not,' in Daly (1992).

(1990) 'Sustainable Development: From Concept and Theory Toward Operational Principles,' in Daly (1992).

(1992) *Steady State Economics*, 2nd edition, London: Earthscan, 1992.

Daly, H. E. and Cobb, J. B.

(1989) *For the Common Good*, London: Green Print, 1990.

Deutscher, M.

(1968) 'Popper's Problem of an Empirical Basis,' *Australasian Journal of Philosophy* **46**, 1968.

Dilworth, C.

(1976) 'Incommensurability and Scientific Progress,' manuscript, 1976.

(1978) 'On the Nature of the Relation Between Successive Scientific Theories,' *Epistemologia* **1**, 1978.

(1992) 'The Linguistic Turn: Shortcut or Detour?,' *Dialectica* **46**, 1992.

(2007) *The Metaphysics of Science*, 2nd ed., Dordrecht: Springer, 2007.

(2008) *The Ecological Predicament of Humankind*, Cambridge: Cambridge University Press, 2008.

Donnellan, K. S.

(1966), 'Reference and Definite Descriptions,' *Philosophical Review* **75**, 1966.

Duhem, P.

(1906) *The Aim and Structure of Physical Theory*, P. P. Wiener tr., N.Y.: Atheneum, 1962.

Edwards, P. ed.

(1967) *Encyclopedia of Philosophy*, N.Y. and London: Macmillan, 1967.

Einstein, A.

(1952) 'Relativity and the Problem of Space,' Appendix V of his *Relativity*, London: Methuen & Co., 1954.

Ellis, B.

(1965) 'The Origin and Nature of Newton's Laws of Motion,' in Colodny ed. (1965).

(1966) *Basic Concepts of Measurement*, Cambridge: Cambridge University Press, 1966.

Elzinga, A.

(1981–1982) Review of the first edition of *Scientific Progress*, *Lychnos*, 1981–1982: 266–268.

Feigl, H. and Maxwell, G. eds.

(1962) *Minnesota Studies in the Philosophy of Science* **III**, Minneapolis: University of Minnesota Press, 1962.

Feyerabend, P. K.

(1962) 'Explanation, Reduction, and Empiricism,' in Feigl & Maxwell eds. (1962).

(1963) 'How to be a Good Empiricist – A Plea for Tolerance in Matters Epistemological,' in P. H. Nidditch ed., *The Philosophy of Science*, London: Oxford University Press, 1968.

(1965a) 'On the "Meaning" of Scientific Terms,' *Journal of Philosophy* **62**, 1965.

(1965b) 'Reply to Criticism,' in R. S. Cohen and M. W. Wartofsky eds., *In Honor of Philipp Frank*, Dordrecht: D. Reidel, 1965.

(1965c) 'Problems of Empiricism,' in Colodny ed. (1965).

(1970) 'Consolations for the Specialist,' in Lakatos & Musgrave eds. (1970).

(1974) 'Popper's *Objective Knowledge*,' *Inquiry* **17**, 1974.

(1975) *Against Method*, London: New Left Books, 1975.

(1977) 'Changing Patterns of Reconstruction,' *British Journal for the Philosophy of Science* **28**, 1977.

(1978) *Science in a Free Society*, London: New Left Books, 1978.

Feyerabend, P. K. and Maxwell, G. eds.

(1966) *Mind, Matter, and Method*, Minneapolis: University of Minnesota Press, 1966.

Flamm, D.

(1983) 'Ludwig Boltzmann and his Influence on Science,' *Studies in the History and Philosophy of Science* **14**, 1983.

Frege, G.

(1879) *Begriffsschrift* (Chapter 1) in Geach & Black eds. (1980).

(1891) 'Funktion und Begriff,' translated as: 'Function and Concept' in Geach & Black eds. (1980).

(1892) 'Über Sinn und Bedeutung,' translated as: 'On Sense and Nominatum' in Copi & Gould eds. (1967).

(1893) *Grundgesetze der Arithmetik*, vol. 1, in Geach & Black eds. (1980).

Fürth, R.

(1969) 'The Role of Models in Theoretical Physics,' in R. S. Cohen and M. W. Wartofsky eds., *Proceedings of the Boston Colloquium for the Philosophy of Science 1966–1968*, Dordrecht: D. Reidel, 1969.

Gahrton, P.

(1988) *Vad Vill De Gröna?*, Gothenburg: Bokförlaget Korpen, 1988.

Galilei, G.

(1638) *Dialogues Concerning Two New Sciences*, H. Crew and A. de Salvio trs., N.Y.: Dover Publications, 1954.

Gärdenfors, P.

(1982) Review of the first edition of *Scientific Progress* (in Swedish), *Filosofisk tidskrift* **3** (1), 1982: 44–46.

Geach, P. and Black, M. eds.

(1980) *Translations from the Philosophical Writings of Gottlob Frege*, Oxford: Basil Blackwell, 1980.

Giedymin, J.

(1970) 'The Paradox of Meaning Variance,' *British Journal for the Philosophy of Science* **21**, 1970.

Golley, F. B.

(1990) 'The Ecological Context of a National Policy of Sustainability,' in B. Aniansson and U. Svedin eds., *Towards an Ecologically Sustainable Economy*, Stockholm: Swedish Council for Planning and Co-ordination of Research, 1990.

Graves, J. C.

(1971) *The Conceptual Foundations of Contemporary Relativity Theory*, Cambridge, Mass.: MIT Press, 1971.

Guthrie, W. K. C.

(1965) *A History of Greek Philosophy*, vol. II, Cambridge: Cambridge University Press, 1965.

Haack, S.

(1976) '"Is it True What They Say About Tarski?",' *Philosophy* **51**, 1976.

Hanson, N. R.

(1958) *Patterns of Discovery*, Cambridge: Cambridge University Press, 1958.

(1966) 'Equivalence: The Paradox of Theoretical Analysis,' in Feyerabend & Maxwell eds. (1966).

Hanson, N. R. et al.

(1970) 'Discussion at the Conference on Correspondence Rules,' in Radner & Winokur eds. (1970).

Harré, R.

(1970a) *The Principles of Scientific Thinking*, Chicago: University of Chicago Press, 1970.

(1970) 'Constraints and Restraints,' *Metaphilosophy* **1**, 1970.

Hempel, C. G.

(1962) 'Deductive-Nomological vs. Statistical Explanation,' in Feigl & Maxwell eds. (1962).

(1965) *Aspects of Scientific Explanation*, N.Y.: Free Press, 1965.

(1966) 'Recent Problems of Induction,' in Colodny ed. (1966).

(1970) 'On the "Standard Conception" of Scientific Theories,' in Radner & Winokur eds. (1970).

Hempel, C. G. and Oppenheim, P.

(1948) 'Studies in the Logic of Explanation,' *Philosophy of Science* **15**, 1948.

Hesse, M.

(1963) 'A New Look at Scientific Explanation,' *Review of Metaphysics* **17**, 1963–1964.

(1966) *Models and Analogies in Science*, Notre Dame, Indiana: University of Notre Dame Press, 1966.

Hirst, R. J.

(1967) 'Primary and Secondary Qualities,' in vol. 6 of Edwards ed. (1967).

Hutten, E. H.

(1954) 'The Rôle of Models in Physics,' *British Journal for the Philosophy of Science* **4**, 1953–1954.

Jansson, AM.

(1988) 'The Ecological Economics of Sustainable Development – Environmental Conservation Reconsidered,' in U. Svedin et al. eds., *Perspectives of Sustainable Development*, Stockholm Studies in Natural Resources Management, No. 1, 1988.

Johansson, I.

(1978) 'Hönan och Galvanometern,' *Några Aktuella Vetenskapsfilosofiska Problem*, Umeå Studies in the Theory and Philosophy of Science, No. 14, 1978.

Jordan, P.

(1944) *Physics of the 20th Century*, E. Oshry tr., N.Y.: Philosophical Library, 1944.

Kaiser, M.

(1979) 'On Sneed and Rationality,' paper presented at the Inter-University Centre, Dubrovnik, April, 1979.

Kant, I.

(1781, 1787) *Critique of Pure Reason,* N. Kemp Smith tr., London and Basingstoke: Macmillan, 1982.

Kenner, L.

(1965) 'The Triviality of the Red-Green Problem,' *Analysis* **25**, 1964–1965.

Keynes, J. M.

(1921) *A Treatise on Probability*, London: Macmillan and Co., 1921.

Kneale, W.

(1964) 'On Popper's Use of the Notion of Absolute Logical Probability,' in M. Bunge ed., *The Critical Approach to Science and Philosophy*, London: Free Press of Glencoe, 1964.

Krajewski, W.

(1977) *Correspondence Principle and Growth of Science*, Dordrecht: D. Reidel, 1977.

Kripke, S.

(1980) *Naming and Necessity*, Oxford: Basil Blackwell, 1980.

Kuhn, T. S.

(1957) *The Copernican Revolution*, Cambridge, Mass.: Harvard University Press, 1957.

(1961) 'The Function of Measurement in Modern Physical Science,' *Isis* **52**, 1961.

(1962) *The Structure of Scientific Revolutions*; vol. 2, no. 2, International Encyclopedia of Unified Science, Chicago: University of Chicago Press, 1970.

(1970a) 'Postscript – 1969,' in Kuhn (1962).

(1970b) 'Logic of Discovery or Psychology of Research?,' in Lakatos & Musgrave eds. (1970).

(1970c) 'Reflections on My Critics,' in Lakatos & Musgrave eds. (1970).

(1971) 'Notes on Lakatos,' in R. C. Buck and R. S. Cohen eds., *PSA 1970: In Memory of Rudolf Carnap*, Urbana: University of Illinois Press, 1974.

(1974) 'Second Thoughts on Paradigms,' in F. Suppe ed., *The Structure of Scientific Theories*, Urbana: University of Illinois Press, 1974.

(1976) 'Theory-Change as Structure-Change: Comments on the Sneed Formalism,' *Erkenntnis* **10**, 1976.

(1977) *The Essential Tension*, Chicago: University of Chicago Press, 1977.

Lakatos, I.

(1968) 'Changes in the Problem of Inductive Logic,' in I. Lakatos ed., *The Problem of Inductive Logic*, Amsterdam: North-Holland, 1968.

(1970) 'Falsification and the Methodology of Scientific Research Programmes,' in Lakatos & Musgrave eds. (1970).

Lakatos, I. and Musgrave, A. eds.

(1970) *Criticism and the Growth of Knowledge*, Cambridge: Cambridge University Press, 1970.

Laudan, L.

(1977) *Progress and Its Problems*, Berkeley: University of California Press, 1977.

Lauener, H.

(1983) Review of the first edition of *Scientific Progress*, *Dialectica* **37**, 1983: 67–69.

Linsky, L.

(1951) 'Reference and Referents,' in Caton ed. (1963).

Locke, J.

(1690) *An Essay Concerning Human Understanding*, N.Y.: Dover Publications, 1959.

Lucretius

On the Nature of the Universe, Harmondsworth: Penguin Books, 1951.

Mach, E.

(1883) *The Science of Mechanics*, Lasalle, Illinois: Open Court, 1960.

Machlup, F.

(1952) *The Economics of Seller's Competition*, Baltimore: Johns Hopkins Press, 1952.

MacNeill, J.

(1989) '"Our Common Future," Sustaining the Momentum,' in F. Archibugi and P. Nijkamp eds., *Economy and Ecology: Towards Sustainable Development*, Dordrecht: Kluwer, 1989.

Martin, M.

(1971) 'Referential Variance and Scientific Objectivity,' *British Journal for the Philosophy of Science* **22**, 1971.

Mates, B.

(1972) *Elementary Logic*, N.Y.: Oxford University Press, 1972.

Mayr, D.

(1976) 'Investigations of the Concept of Reduction I,' *Erkenntnis* **10**, 1976.

McKinsey, J. C. C., Sugar, A. C. and Suppes, P.

(1953) 'Axiomatic Foundations of Classical Particle Mechanics,' *Journal of Rational Mechanics and Analysis* **2**, 1953.

McMullin, E.

(1984) 'Two Ideals of Explanation in Natural Science,' in P. A. French et al. eds., *Midwest Studies in Philosophy*, vol. **IX**, Minneapolis: University of Minnesota Press, 1984.

Meadows, D. H. et al.

(1972) *The Limits to Growth*, London: Earth Island, 1972.

Miller, D. W.

(1974) 'Popper's Qualitative Theory of Verisimilitude,' *British Journal for the Philosophy of Science* **25**, 1974.

Mitton, R. G.

(1939) *Heat*, London: J. M. Dent and Sons, 1945.

Moon, M. ed.

(1989) *Gröna Läseboken*, Lund: Miljöpartiet de Gröna, 1989.

Moulines, C. U.

(1976) 'Approximate Application of Empirical Theories: A General Explication,' *Erkenntnis* **10**, 1976.

Nagel, E.

(1961) *The Structure of Science*, N.Y.: Harcourt, Brace & World, 1961.

Newton, I.

(1687) *Mathematical Principles of Natural Philosophy*, A. Motte tr., Berkeley: University of California Press, 1947.

Niles, T. M. T.

(1999) 'A Better Economy the World Over,' *San Francisco Chronicle*, 29 November 1999.

Nowak, L.

(1979) 'Idealization and Rationalization,' *Epistemologia* **2**, Special Issue, 1979.

(1980) *The Structure of Idealization*, Dordrecht: D. Reidel, 1980.

O'Connor, D. J.

(1955) 'Incompatible Properties,' *Analysis* **15**, 1955.

Page, T.

(1991) 'Sustainability and the Problem of Valuation,' in Costanza ed. (1991).

Partington, J. R.

(1961) *A History of Chemistry*, vol. 2, London: Macmillan and Co., 1961.

Phillips, R.

(1983) Review of the first edition of *Scientific Progress*, *Australasian Journal of Philosophy* **61**, 1983: 213–216.

Poincaré, H.

(1902) *Science and Hypothesis*, N.Y.: Dover Publications, 1952.

Popper, K. R.

(1934) *The Logic of Scientific Discovery*, London: Hutchinson & Co., 1959.

(1949) 'The Bucket and the Searchlight: Two Theories of Knowledge,' in Popper (1973).

(1957a) 'The Aim of Science,' in Popper (1973).

(1957b) 'Science: Conjectures and Refutations,' in Popper (1962).

(1958) 'On the Status of Science and Metaphysics,' in Popper (1962).

(1959) 'New Appendices' (and new footnotes), in Popper (1934).

(1962) *Conjectures and Refutations*, N.Y.: Harper & Row, 1963.

(1972) *Objective Knowledge*, Oxford: Oxford University Press, 1972.

(1973) *Objective Knowledge* (with corrections), Oxford: Oxford University Press, 1973.

(1975) 'The Rationality of Scientific Revolutions,' in R. Harré ed., *Problems of Scientific Revolution*, Oxford: Clarendon Press, 1975.

Porritt, J.

(1992) 'Sustainable Development: Panacea, Platitude or Downright Deception?,' *Energy and the Environment: The Linacre Lectures*, Oxford: Oxford University Press, 1992.

Post, H. R.

(1971) 'Correspondence, Invariance and Heuristics,' *Studies in History and Philosophy of Science* **2**, 1971.

Putnam, H.

(1962) 'What Theories Are Not,' in E. Nagel et al. eds., *Logic, Methodology and Philosophy of Science*, Stanford: Stanford University Press, 1962.

Quine, W. v. O.

(1953) 'On What There Is,' in his *From a Logical Point of View*, N.Y.: Harper Torchbooks, 1961.

(1960) *Word and Object*, Cambridge, Mass.: MIT Press, 1960.

Radner, M. and Winokur, S. eds.

(1970) *Minnesota Studies in the Philosophy of Science* **IV**, Minneapolis: University of Minnesota Press, 1970.

Ramsey, F. P.

(1931) 'Theories,' in his *The Foundations of Mathematics*, London: Kegan Paul, Trench, Trubner & Co., 1931.

Reichenbach, H.

(1928) *The Philosophy of Space and Time*, N.Y.: Dover Publications, 1958.

Robinson, J. M.

(1968) *An Introduction to Early Greek Philosophy*, Boston: Houghton Mifflin Company, 1968.

Røpke, I.

(1992) 'Trade Development and Sustainability – A Critical Assessment of the "Free Trade Dogma",' paper delivered at the International Society of Environmental Economics meeting in Stockholm, 1992.

Rossi, P. A.

(1982) Review of the first edition of *Scientific Progress*, *Epistemologia* **5**, 1982: 186–188.

Russell, B.

(1905) 'On Denoting,' in Copi & Gould eds. (1967).

(1940) *An Inquiry into Meaning and Truth*, N.Y.: W. W. Norton & Co., 1940.

Ryle, G.

(1953) 'Ordinary Language,' in Caton ed. (1963).

Schirn, M. ed.

(1976) *Studies on Frege II: Logic and Philosophy of Language*, Stuttgart-Bad: Frommann-Holzboog, 1976.

Schumacher, E. F.

(1973) *Small is Beautiful*, N.Y.: Harper & Row, 1973.

Scriven, M.

(1962) 'Explanations, Predictions, and Laws,' in Feigl & Maxwell eds. (1962).

Shapere, D.

(1966) 'Meaning and Scientific Change,' in Colodny ed. (1966).

Shrader, D.

(1985) Review of the first edition of *Scientific Progress*, *British Journal for the Philosophy of Science* **36**, 1985: 221–225.

Sneed, J. D.

(1971) *The Logical Structure of Mathematical Physics*, Dordrecht: D. Reidel, 1979.

(1976) 'Philosophical Problems in the Empirical Science of Science: A Formal Approach,' *Erkenntnis* **10**, 1976.

Spector, M.

(1965) 'Models and Theories,' *British Journal for the Philosophy of Science* **16**, 1965–1966.

280

Stegmüller, W.

(1973) *The Structure and Dynamics of Theories*, N.Y.: Springer-Verlag, 1976.

(1976) 'Accidental ('Non-Substantial') Theory Change and Theory Dislodgment: To What Extent Logic Can Contribute to a Better Understanding of Certain Phenomena in the Dynamics of Theories,' *Erkenntnis* **10**, 1976.

(1979) *The Structuralist View of Theories*, Berlin: Springer-Verlag, 1979.

Strawson, P. F.

(1950) 'On Referring,' in Copi & Gould eds. (1967).

(1952) 'Logical Theory,' in Copi & Gould eds. (1967).

Suppes, P.

(1967) *Set-Theoretical Structures in Science*, stencil, Stanford University, Stanford, 1970.

Tichý, P.

(1974) 'On Popper's Definitions of Verisimilitude,' *British Journal for the Philosophy of Science* **25**, 1974.

Ullman, S.

(1962) *Semantics*, Oxford: Basil Blackwell, 1967.

von Wright, G. H.

(1986) *Vetenskapen och Förnuftet*, Stockholm: MånPocket, 1986.

Wallace, W. A.

(1974) *Causality and Scientific Explanation*, vol. 2, Ann Arbor: University of Michigan Press, 1974.

Ward, B.

(1979) *The Ideal Worlds of Economics*, N.Y.: Basic Books, 1979.

Wartofsky, M. W.

(1968) *Conceptual Foundations of Scientific Thought*, N.Y.: Macmillan Co., 1968.

Welin, S.

(1982) 'Vetenskapliga framsteg?,' *Filosofisk tidskrift* **3** (3), 1982.

Wikstrom, J. H. and Alston, R. M.

(1992) 'Is Economics Lost in a Scientific Revolution?,' paper delivered at the International Society of Environmental Economics meeting in Stockholm, 1992.

Wilkinson, R. G.

(1973) *Poverty and Progress. An Ecological Perspective on Economic Development*, N.Y.: Praeger Publishers, 1973.

Wittgenstein, L.

(1921) *Tractatus Logico-Philosophicus*, D. F. Pears and B. F. McGuinness trs., London: Routledge & Kegan Paul, 1961.

(1953) *Philosophical Investigations*, G. E. M. Anscombe tr., Oxford: Basil Blackwell, 1972.

Woodward, J.

(1979) 'Scientific Explanation,' *British Journal for the Philosophy of Science* **30**, 1979.

World Commission on Environment and Development

(1987) *Our Common Future*, Oxford and N.Y.: Oxford University Press, 1987.

INDEX